INTERIOR FINISH:

MORE TRICKS OF THE TRADE

BOB SYVANEN

The East Woods Press

LIBRARY OF CONGRESS CATALOGING IN PUBLICATION DATA

SYVANEN, BOB, 1928-
        INTERIOR FINISH: MORE TRICKS OF THE TRADE FROM AN OLD-STYLE CARPENTER.

        1. CARPENTRY- AMATEURS' MANUALS.   I. TITLE.
TH5606.S96 1982   694'.6   82-8907
ISBN 0-914788-56-6

PRINTED IN THE UNITED STATES OF AMERICA

THIS BOOK WAS FIRST PRINTED BY THE AUTHOR IN 1981.
IF YOU HAVE ANY CARPENTRY QUESTIONS, INCLUDE A STAMPED, SELF-ADDRESSED ENVELOPE AND SEND THEM TO:
BOB SYVANEN
179 UNDERPASS ROAD
BREWSTER, MASSACHUSETTS 02631

AN EAST WOODS PRESS BOOK
FAST AND McMILLAN PUBLISHERS, INC.
429 EAST BOULEVARD
CHARLOTTE, NC 28203

INTRODUCTION

THIS BOOK WAS DONE WITH THE SAME INTENT AS WERE THE CARPENTRY AND DRAFTING BOOKS: TO HELP AND ENCOURAGE.

INTERIOR FINISHING HAS AS MANY UNFORESEEN PROBLEMS AS THE FRAMING. UNFORTUNATELY FINISH WORK IS VERY VISIBLE, BUT KEEP IN MIND THAT IT IS A HOUSE YOU ARE BUILDING AND NOT A PIANO.

THINGS ARE ALWAYS HAPPENING ON A JOB THAT YOU HADN'T PLANNED ON, BUT THE SOLUTION IS THERE. SOMETIMES IT'S," OH YEAH, I DID SOMETHING LIKE THAT BACK IN 1945" AND OTHER TIMES IT'S BRAND NEW (TO YOU). LET THAT INNER WISDOM FIND IT FOR YOU. A LITTLE EXPERIENCE CAN'T HURT EITHER AND THAT'S WHERE I HOPE I CAN HELP.

BOB SYVANEN
BREWSTER, MASSACHUSETTS
1981

TRANSLATION:
HO BOY, THE TRILOGY IS FINISHED.

4

CONTENTS

THE EXTERIOR IS FINISHED. NOW COMES THE INTERIOR.

OF COURSE WE TIMED IT SO THAT THE BAD WEATHER WILL BE SPENT DOING THE INSIDE WORK.

I DO NOT MENTION SAFETY PRACTICES OR DEVICES, ASSUMING THAT BASIC PROCEDURES WOULD BE PRACTICED BY THE READER. SAFETY GOGGLES, PROTECTIVE CLOTHING, PROPER VENTILATION, AND SAFE LADDERS, ARE ALL PART OF GOOD CARPENTRY. OSHA RULES, BUILDING CODES, MANUFACTURERS' INSTRUCTIONS, AND, ABOVE ALL, GOOD OLD COMMON SENSE, SHOULD BE CONSIDERED IN ALL CARPENTRY JOBS.

THERE IS A LOT OF BLOCKING TO GO IN
BEFORE THE FINISHED CEILING AND
FINISHED WALLS GO ON.

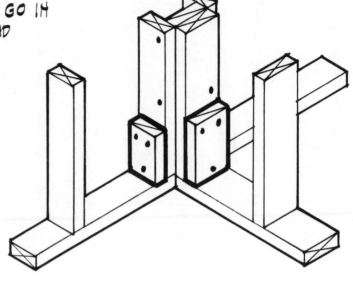

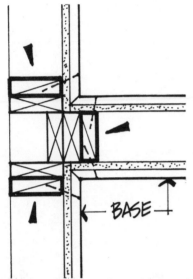

PLAN @ INSIDE CORNER

THE INSIDE CORNERS NEED BLOCKING TO MAKE NAIL-
ING THE BASEBOARD EASIER. YOU WON'T HAVE TO
REACH INTO THE CORNER TO FIND NAILING.

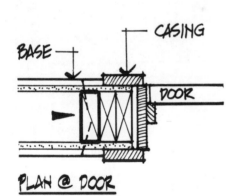

PLAN @ DOOR

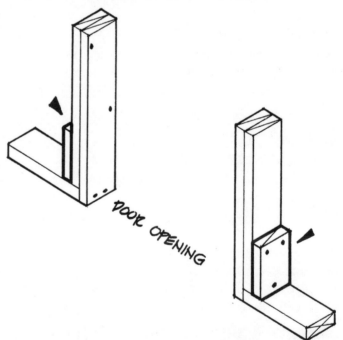

DOOR OPENING

BLOCKING AT DOOR OPENINGS IS REQUIRED FOR THE SAME REASON. THE
CASING USUALLY EXTENDS PAST THE JAMB STUDS, LEAVING VERY LITTLE
TO NAIL TO.

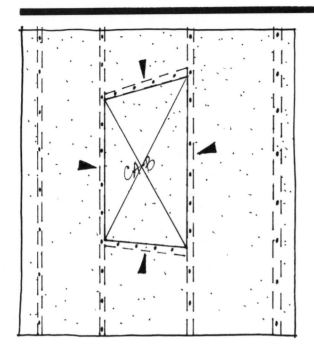

MEDICINE CABINETS ARE SCREWED TO SIDE BLOCKING, SO MAKE SURE THERE IS SOMETHING TO SCREW INTO. THE TOP AND BOTTOM SHOULD BE BLOCKED TOO. IT HELPS TO MARK ITS LOCATION AS THE SHEETROCK GOES ON, BUT EVEN UNMARKED IT IS EASY TO LOCATE.

1X6 BLOCKING IS GOOD ENOUGH FOR TOWEL BARS, SHOWER HEADS AND FAUCETS. IT'S A GOOD IDEA TO MARK THESE LOCATIONS ON THE SHEETROCK.

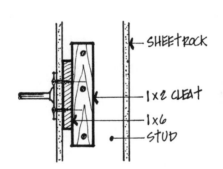

SHEETROCK

1X2 CLEAT

1X6

STUD

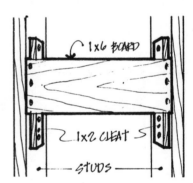

1X6 BOARD

1X2 CLEAT

STUDS

PANELING REQUIRES SOME-THING MORE SUBSTANTIAL.

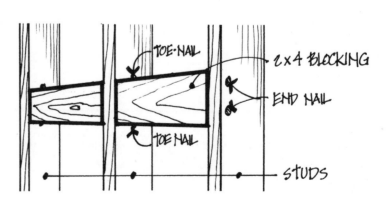

TOE-NAIL

2X4 BLOCKING

END NAIL

TOE-NAIL

STUDS

9

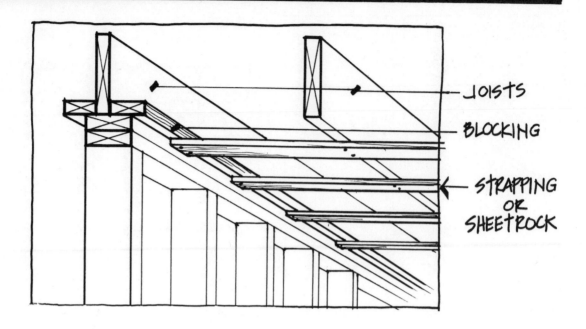

JOISTS

BLOCKING

STRAPPING OR SHEETROCK

THE CEILING SHOULD BE PREPARED FOR EITHER STRAPPING OR SHEETROCK. EITHER WAY, IT WILL NEED BLOCKING.

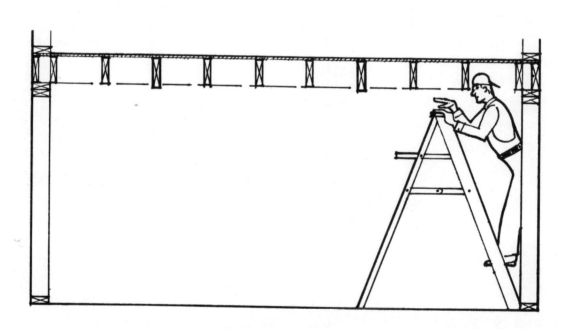

IF THE SHEETROCK IS TO BE INSTALLED DIRECTLY ON THE JOISTS, EYEBALL DOWN TO SEE IF THERE ARE ANY LOW-HANGING ONES. IF ANY NEED TRIMMING, SNAP A CHALK LINE AND CUT WITH A SKIL-SAW OR HATCHET.

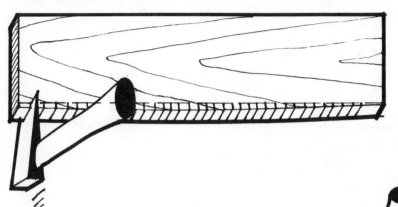

IF YOU CAN'T CUT IT WITH A SKIL-SAW, A SERIES OF CUTS ACROSS THE BOTTOM EDGE OF THE JOIST WITH A SHARP HATCHET......

FOLLOWED BY STROKES PARALLEL TO AND AT THE CHALK LINE WILL DO A QUICK AND NEAT JOB.

A "STRONGBACK" MIGHT BE ALL THAT IS NEEDED TO EVEN UP THE BOTTOMS, BUT NOT THIS KIND OF STRONG BACK.

2×6

2×4

CEILING JOIST

A STRONGBACK IS A 2×6 NAILED TO A 2×4 AND THEN NAILED TO THE TOPS OF THE CEILING JOISTS, PULLING ANY BOWED ONES INTO LINE. A 2×8 CAN BE USED FOR MORE STRENGTH.

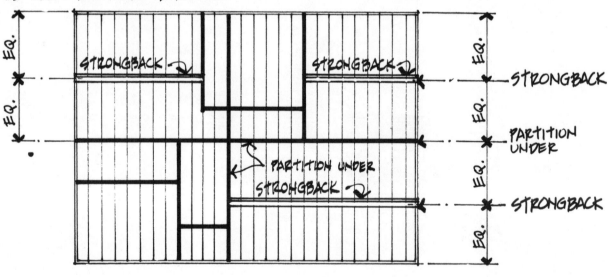

STRONGBACK

STRONGBACK

STRONGBACK

PARTITION UNDER

PARTITION UNDER

STRONGBACK

STRONGBACK

EQ.

CEILING JOIST PLAN

THE STRONGBACK IS LOCATED AT MIDSPAN AND ONLY WHERE THE FLOOR SPACE ABOVE THE JOISTS IS NOT USABLE.

12

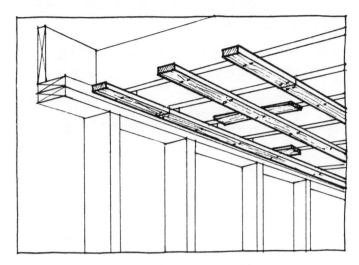

NEW ENGLAND IS THE ONLY PLACE I KNOW OF THAT USES 1 x 3 STRAPPING AS A BASE FOR A SHEETROCK CEILING. IT IS A TERRIFIC WAY TO GET A FLAT CEILING, BUT MOST CARPENTERS JUST STRAP THE CEILING AND THEN INSTALL THE SHEETROCK WITHOUT LEVELING IT.

EYEBALL FOR ANY BAD JOISTS AND FIX BEFORE STRAPPING. THE EASIEST WAY TO GET THE JOB DONE IS TO NAIL UP ALL THE STRAPPING AND THEN EYE-BALL FOR ADJUSTING. FOR A SUPER JOB WORK TO A STRETCHED STRING.

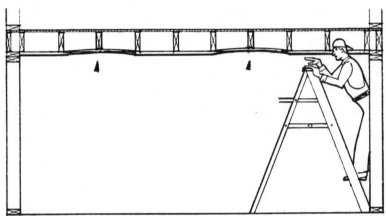

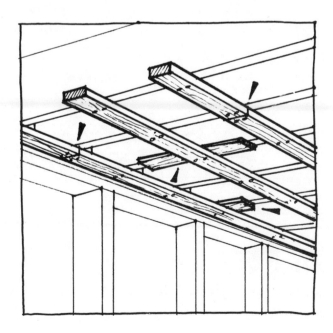

USE SHINGLE TIPS, BETWEEN STRAP-PING AND JOISTS, TO ADJUST. I LIKE TO DOUBLE NAIL FOR A MORE STABLE CONDITION, BUT A SINGLE NAIL WILL SURELY HOLD. WHEN A CEILING NEEDS JOINTS IN THE STRAPPING, ALTER-NATE THEM.

PREPARATION

MEASURE THE VARIOUS
SIZE STRAPPING LENGTHS
AND CUT THEM WHILE
THE BUNDLE IS STILL
TIED TOGETHER.

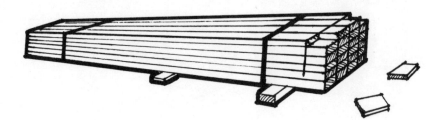

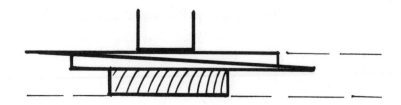

TWO SHINGLES USED BUTT
TO TIP MAKES A BETTER
SHIM JOB.

ONE SHINGLE WILL TIP
THE FACE. IF THE FACE
IS ALREADY TIPPED, THEN
ONE SHINGLE CAN CORRECT
IT.

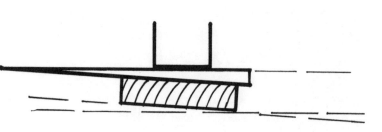

IF PLASTER IS TO BE USED THEN
THERE IS NO NEED TO WORRY
ABOUT THE JOIST BOTTOMS; THE
PLASTER WILL TAKE CARE OF THAT.
IN THE OLD DAYS, THE BOSS
WOULD ENCOURAGE AN APPRENTICE
TO PUT ON A THICK COAT OF
PLASTER BY THROWING A HAND-
FUL OF SAND INTO THE SKIMPY
COAT ALREADY APPLIED.

LETS GET SOME
PLASTER ON THERE
DAVE !!!!

CHECK THE WALLS FOR BADLY
BOWED STUDS BY HOLDING A
LONG STRAIGHT 2x4 AGAINST
THE WALL. IT WILL BE OB-
VIOUS WHICH ONES ARE BAD.
A BADLY BOWED STUD IN A
FINISHED WALL REALLY SHOWS,
SO IT IS WISE TO REPLACE
OR STRAIGHTEN IT.

TO STRAIGHTEN A BOWED STUD, CUT WELL INTO IT ON
THE SIDE OPPOSITE THE HUMP.

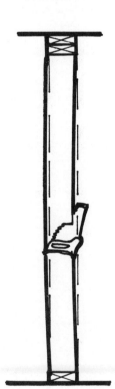

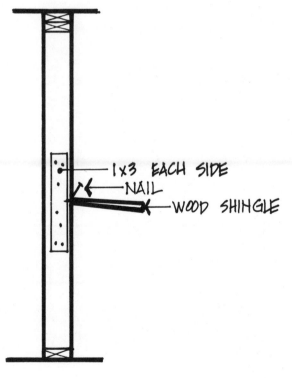

1x3 EACH SIDE
NAIL
WOOD SHINGLE

THE STUD CAN THEN BE PUSHED TO A
STRAIGHT POSITION. WOOD SHINGLE TIPS
DRIVEN INTO THE CUT WILL KEEP THE
STUD STRAIGHT WHILE CLEATS ARE
NAILED ON EACH SIDE. IT MIGHT TAKE
TWO SUCH CUTS IF THE BOW IS BAD

15

INSULATION

INSULATION IS A VERY IMPORTANT JOB IN FINISHING A HOUSE, PARTICULARLY IN THESE DAYS OF HIGH FUEL COSTS. THERE ARE FIVE PROBLEM AREAS THAT SHOULD BE DONE WITH CARE: CORNERS, PARTITION INTERSECTIONS, ROOFS, SPACE BEHIND ELECTRICAL OUTLET BOXES, AND VAPOR BARRIERS.

THE STANDARD OUTSIDE CORNER DOES NOT INSULATE WELL.

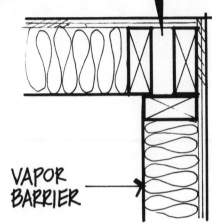

VAPOR BARRIER

PLAN

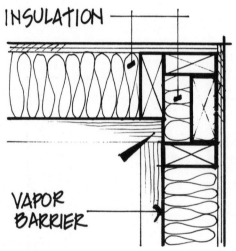

INSULATION

VAPOR BARRIER

PLAN

THIS CORNER ALLOWS FOR GOOD INSULATION.

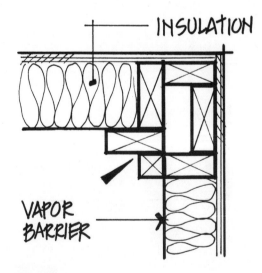

INSULATION

VAPOR BARRIER

PLAN

IF THE "FINNISH WALL" (SCANDINAVIAN) IS USED AND THE 2x2's ARE PLANTED ON THE FACE OF THE STUDS AND PARALLEL, INSULATE BEFORE THE 2x2's GO ON IN THE CORNER. IF THE 2x2's ARE PERPENDICULAR TO THE STUDS, THERE IS NO PROBLEM.

16

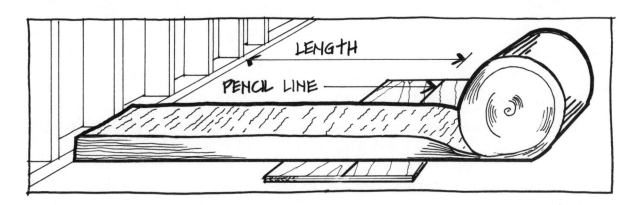

A PENCIL LINE ON A PIECE SCRAP PLYWOOD IS A QUICK, EASY GUIDE FOR DUPLICATE CUTTING OF INSULATION. MOVE THE PLYWOOD WITH THE LINE TO THE LENGTH OF THE INSULATION TO BE CUT.

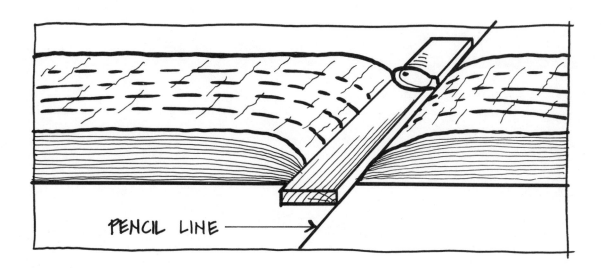

IF THERE IS A PAPER BACKING ON THE INSULATION IT SHOULD BE DOWN WITH THE FLUFF SIDE UP. COMPRESS THE INSULATION WITH A BOARD THAT LINES UP WITH THE PENCIL LINE AND CUT WITH A SHARP UTILITY KNIFE.

INSULATION

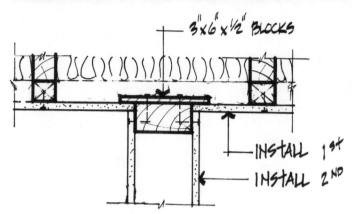

3"x6"x½" BLOCKS

INSTALL 1ST

INSTALL 2ND

WHERE A PARTITION BUTTS AN EXTERIOR WALL, I FAVOR SHEETROCK CLIPS OR 3"x6"x½" PLYWOOD BLOCKS AT 16" O.C. THIS ALLOWS FOR UNBROKEN INSULATION BETWEEN STUDS.

INSIDE CORNER PLAN

THERE MUST BE AN AIR SPACE BETWEEN THE TOP OF THE INSULATION AND THE BOTTOM OF THE ROOFING BOARDS TO PREVENT CONDENSATION.

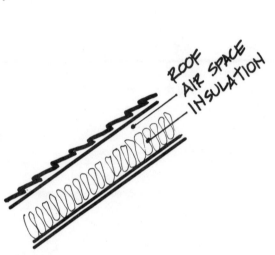

ROOF
AIR SPACE
INSULATION

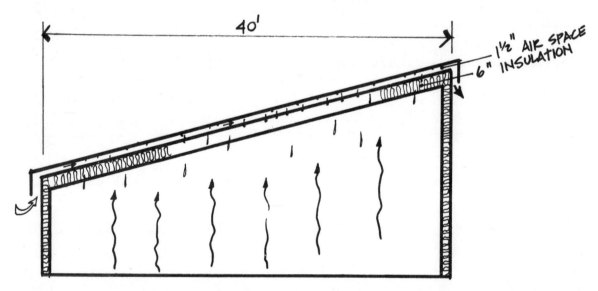

40'

1½" AIR SPACE

6" INSULATION

I HAVE THIS CONDITION IN MY HOUSE AND IT "RAINS" INSIDE BECAUSE NOT ENOUGH AIR MOVES IN THE 40 FOOT ROOF. THE ROOF IS COLD AND WHEN THE WARM AIR HITS IT, YOU WOULDN'T BELIEVE THE CONDENSATION.

18

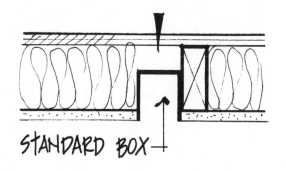

STANDARD BOX →

AN ELECTRICAL OUTLET BOX PRETTY MUCH FILLS THE SPACE IN A 2x4 STUD WALL. THE RESULT IS THAT DIRECTLY BEHIND THE BOX THERE IS NO INSULATION.

THE FINNISH WALL ELIMINATES THIS PROBLEM BY KEEPING THE ELECTRICAL WORK IN THE 1½" AIR SPACE. THIS IS NOT WITHOUT ITS PROBLEMS BECAUSE OF THE SHALLOW OUTLET BOXES THAT MUST BE USED. THE SHALLOW BOX GETS CROWDED IN A HURRY, BUT JUNCTION BOXES WILL HELP.

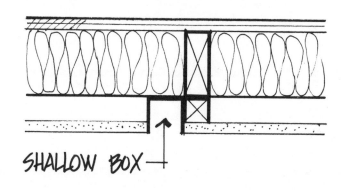

SHALLOW BOX →

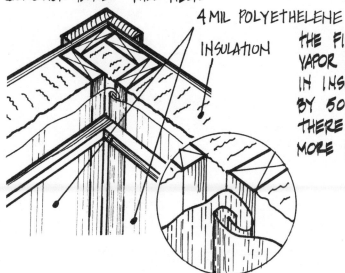

4 MIL POLYETHELENE

INSULATION

THE FINNISH WALL ALSO TAKES CARE OF THE VAPOR BARRIER PROBLEM. 3% MOISTURE IN INSULATION REDUCES THE "R" VALUE BY 50%, SO THE FEWER BREAKS THERE ARE IN THE VAPOR BARRIER, THE MORE EFFICIENT THE INSULATION WILL BE.

WHEN SHEETS OF POLYETHELENE ARE JOINED, IT IS BEST TO LOCK THE SEAM FOR A GOOD SEAL.

A BASEMENT WALL IS BEST INSULATED ON THE OUTSIDE.

UNDERLAYMENT

THE KITCHEN AND BATHROOM FINISHED FLOOR IS USUALLY VINYL OVER PLYWOOD UNDERLAYMENT. VERSA-BOARD OR ANY COMPOSITION BOARD IS NOT A GOOD PRODUCT TO USE WHERE THERE IS MOISTURE SINCE IT SWELLS WHEN WET. USE PLYWOOD, PLUGGED AND SANDED ONE SIDE; IT'S MADE FOR UNDERLAYMENT.

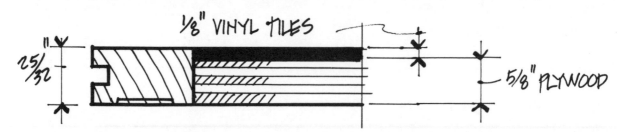

$25/32$" $1/8$" VINYL TILES $5/8$" PLYWOOD

WHEN KITCHEN OR BATHROOM MEETS A HARDWOOD FLOOR, $5/8$" PLYWOOD WITH $1/8$" TILE WILL MAKE BOTH FLOORS ABOUT EVEN.

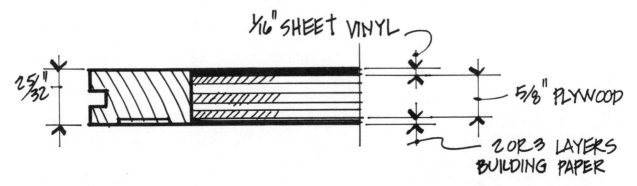

$25/32$" $1/16$" SHEET VINYL $5/8$" PLYWOOD

2 OR 3 LAYERS BUILDING PAPER

IF $1/16$" SHEET VINYL IS USED, A FEW LAYERS OF BUILDING PAPER WILL BRING THE VINYL SURFACE FLUSH WITH THE HARDWOOD. TRY A SAMPLE OF THE PLYWOOD, BUILDING PAPER, AND VINYL AGAINST A PIECE OF FLOORING. CHECK THE PLYWOOD; IT'S APT TO MEASURE ANYTHING THESE DAYS.

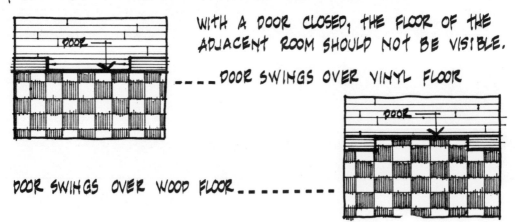

WITH A DOOR CLOSED, THE FLOOR OF THE ADJACENT ROOM SHOULD NOT BE VISIBLE.

----- DOOR SWINGS OVER VINYL FLOOR

DOOR SWINGS OVER WOOD FLOOR-----------

20

IF THE VINYL FLOORING EXTENDS AROUND THE DOOR JAMBS, RUN THE UNDERLAYMENT IN ONE PIECE. THERE IS A LOT OF TRAFFIC IN THESE AREAS AND THE MORE SECURE THE UNDERLAYMENT IS, THE LESS WEAR THERE WILL BE ON THE VINYL. WHERE THE UNDERLAYMENT ENDS AT A DOOR, PROTECT THE EDGE WITH A PIECE OF PLYWOOD NAILED AGAINST IT. IT'S TEMPORARY.

TEMPORARY PROTECTION FOR EDGE OF UNDERLAYMENT.

15# FELT →

CAN BE A SEPARATE PIECE, BUT THIS WAY IS BETTER.

SAME THICKNESS AS UNDERLAYMENT

15# FELT

UNDERLAYMENT

SPIRAL NAIL

SNAP A CHALK LINE ON THE UNDERLAYMENT OVER EVERY JOIST. THE UNDERLAYMENT SHOULD BE NAILED THROUGH THE SUB-FLOOR INTO THE JOISTS FOR A SECURE FLOOR. USE SPIRAL FLOORING NAILS AT THE JOISTS AND RING SHANK NAILS IN BETWEEN AND AROUND THE PERIMETER. IF THE UNDERLAYMENT HAS ANY MOVEMENT, THE NAILS WANT TO WALK OUT AND PUSH THROUGH THE VINYL, SO MAKE SURE IT IS SECURELY NAILED.

SUB FLOOR

FLOOR JOIST

RING SHANK NAIL

SPIRAL FLOORING NAIL

BOTH OF THESE NAILS HAVE GOOD HOLDING POWER

SHEETROCK

THE CEILING IS THE PLACE TO START AND YOU NEED TWO PEOPLE TO DO IT. I HAVE DONE THE JOB ALONE, BUT I DON'T RECOMMEND IT.

THIS GUY IS DESTINED FOR DISASTER.

A COUPLE OF "T" BRACES WILL MAKE THE JOB EASIER. BE SURE YOU CAN REACH THEM AND STILL CONTROL THE SHEET OVERHEAD. IT ALSO HELPS TO STICK A FEW NAILS IN THE SHEETROCK WHERE THE NAILING WILL BE. HOLDING THE SHEET OVERHEAD WHILE FISHING FOR NAILS CAN BE TOUGH.

THE "T" BRACE SHOULD BE A LITTLE LONGER THAN THE FLOOR TO CEILING HEIGHT SO THAT THERE IS A SLIGHT WEDGE FIT. IF IT IS TOO LONG, THE BRACE WON'T STAY IN PLACE AND IF TOO SHORT, THE WHOLE BUSINESS COMES DOWN.

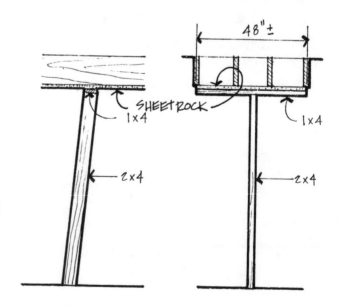

THE EASIEST WAY TO HOLD A SHEET AGAINST THE CEILING IS WITH YOUR HEAD. THE FLATTER THE HEAD THE BETTER.

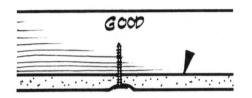

GOOD

WHEN NAILING, MAKE SURE THE SHEET IS PUSHED HARD AGAINST THE JOISTS OR STRAPPING BEFORE DRIVING THE NAIL HOME.

IF YOU DON'T, THE NAIL WILL PULL THROUGH THE SURFACE OF THE SHEETROCK.

BAD

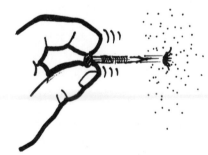

IF A NAIL MISSES A JOIST OR STUD, PULL IT OUT.

HIT THE HOLE WITH A HAMMER HARD ENOUGH TO DEPRESS THE SURFACE WITHOUT BREAKING THE PAPER.

THESE DENTS, OR DIMPLES, CAN BE FILLED EASILY WITH JOINT COMPOUND.

SHEETROCK

IF THE CEILING REQUIRES A BUTTED JOINT AT THE END OF A ROW, IT SHOULD NOT LAND ON A JOIST OR STRAPPING. THE END IS NOT TAPERED AND WHEN TAPED AND SPACKLED IT WILL SHOW A BAD BULGE.

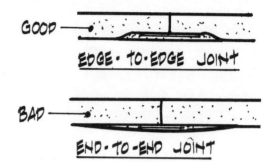

GOOD

EDGE-TO-EDGE JOINT

BAD

END-TO-END JOINT

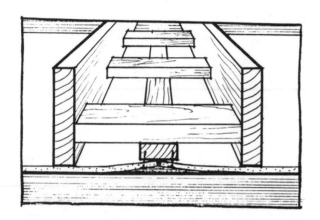

ONE SOLUTION IS TO NAIL UP BLOCKING AT 16" O.C. AND THEN A 2×4 OR STRAPPING DOWN THE MIDDLE, PARALLEL TO THE JOISTS, SO THAT THE SHEETROCK WILL BE DEPRESSED 1/8". THE SHEETROCK EDGES ARE THEN NAILED OR SCREWED TO THE 2×4 OR STRAPPING.

ANOTHER, MORE COMMON WAY IS TO CUT FOUR-12"×12" SQUARES OF SHEETROCK.....

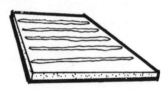

BUTTER THESE PIECES WITH JOINT COMPOUND AND SLIP THEM IN ON THE BACK SIDE OF THE PANEL ALREADY IN PLACE.

THE NEXT SHEET IS NAILED IN PLACE AND A PIECE OF STRAPPING IS PLACED ALONG THE SEAM AND HELD IN PLACE WITH CROSS PIECES OF STRAPPING. THESE CROSS PIECES WILL DEPRESS THE JOINT, AND THE BUTTERED 12"×12"'S WILL DRY HOLDING EVERYTHING IN PLACE. THE JOINT IS THEN TAPED AND SPACKLED LIKE ANY OTHER.

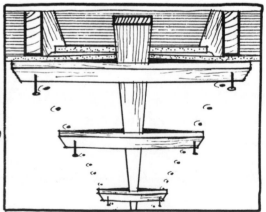

24

WORKING FROM A LADDER OR HORSES IS ALL RIGHT, BUT A SIMPLE PAIR OF STILTS CAN BE A GREAT HELP. I'VE HEARD THAT SOME PLACES HAVE OUTLAWED THEIR USE BECAUSE SOMEONE GOT HURT, SO BE CAUTIOUS. TRY SOME OF YOUR OWN DESIGN AND PRATICE A LITTLE. WHO KNOWS, MAYBE BARNUM AND BAILEY BECKONS.

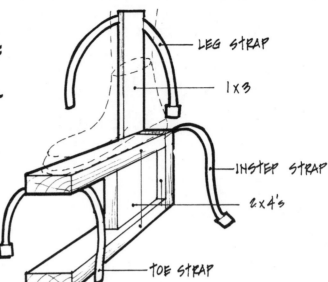

LEG STRAP
1x3
INSTEP STRAP
2x4's
TOE STRAP

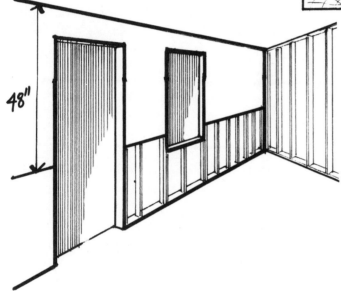

48"

SHEETROCK WALLS ARE PRETTY SIMPLE. THE PANEL THAT TOUCHES THE CEILING SHOULD GO IN FIRST TO INSURE A GOOD JOINT AT THE CEILING.

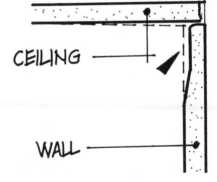

CEILING

WALL

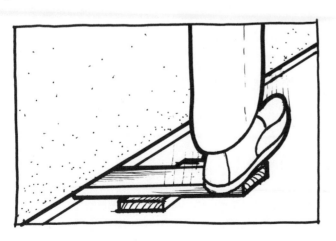

LEAVE ABOUT A HALF INCH GAP AT THE FLOOR SO THAT A WEDGE CAN BE SLIPPED UNDER TO PUSH THE LOWER SHEET UP TIGHT.

SHEETROCK

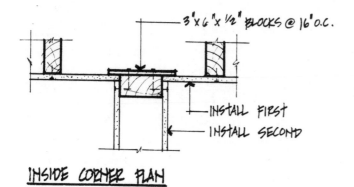

3"x 6"x ½" BLOCKS @ 16" O.C.

INSTALL FIRST
INSTALL SECOND

INSIDE CORNER PLAN

IF CORNER CLIPS OR PLYWOOD BLOCKS ARE USED AT INSIDE CORNERS, INSTALL THE SHEET THAT RUNS PARALLEL TO THE BLOCKS FIRST AND NO NAILING INTO THESE BLOCKS OR CLIPS IS REQUIRED. THE ADJACENT SHEET IS NAILED TO THE STUD IN THE CORNER AND HOLDS THE FIRST IN PLACE. TAPE AND SPACKLE HOLD THE WHOLE THING TOGETHER.

THE ONLY THING TO REMEMBER ABOUT THE OUTSIDE CORNER IS TO MAKE SURE THE SHEETROCK EXTENDS ENOUGH TO THE CORNER THAT THE CORNER BEAD WILL HAVE BACKING BEHIND IT.

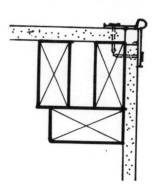

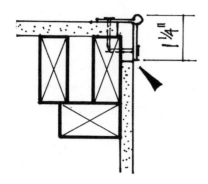

1 ¼"

POOR NAILING CONDITION HERE.

WHEN SHEETROCKING AROUND DOORS AND WINDOWS, RUN THE SHEET BY IN ONE PIECE AND CUT IT OUT FOR THE OPENING. IF PIECES ARE PUT IN OVER THE DOOR OR WINDOW, IT WILL CRACK AT THE SEAM.

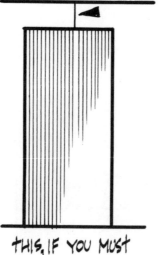

THIS, IF YOU MUST

NEVER THIS

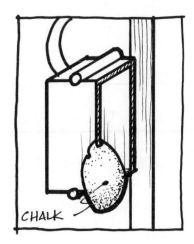

CHALK

SOME PROFESSIONAL SHEETROCKERS COVER THE WALL AND THEN CUT THE OUTLET AND SWITCH HOLES. A SAFER WAY IS TO RUB THE OUTLET BOX WITH BLOCK CHALK, HOLD THE SHEET IN PLACE, SMACK IT WITH AN OPEN PALM AND CUT THE RESULTING MARK LEFT ON THE BACK SIDE OF THE SHEET. ANOTHER WAY IS TO MEASURE THE LOCATION AND MARK IT ON THE SHEET. NOT BAD, BUT BE PREPARED FOR A FEW MISTAKES.

A STANLEY SURFORM IS A GOOD TOOL FOR SHAVING A SHEETROCK EDGE. A PIECE OF EXPANDED WIRE LATH WRAPPED AROUND A BLOCK OF WOOD WORKS JUST AS WELL.

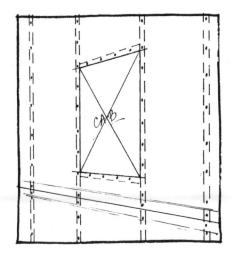

DON'T FORGET TO MARK ALL OPENINGS FOR MEDICINE CABINETS, FANS, ETC., AS THE SHEETS GO UP. THEY ARE EASILY LOST.

AN ALUMINUM "HAWK" CAN BE BOUGHT FOR AROUND $7.00. A PLYWOOD ONE WILL WORK ALMOST AS WELL AND A LOT CHEAPER.

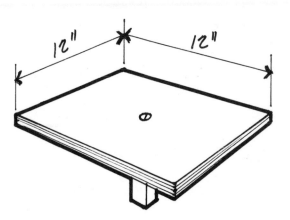

12"   12"

SHEETROCK

A 60 POUND BUCKET OF READY MIX JOINT COMPOUND IS VERY CONVENIENT TO WORK WITH. IT'S EASY TO SCOOP THE COMPOUND OUT OF.

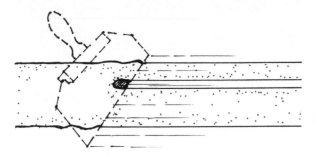

CLEANLINESS IS A MUST. ANY HARD LUMPS OR PIECES OF DIRT WILL MESS UP A JOINT, SO KEEP THE COVER ON.

LOAD UP THE HAWK BY SCOOPING COMPOUND FROM THE BUCKET WITH A WOOD SHINGLE. KEEP THE SHINGLE IN THE BUCKET AND COVER TO KEEP MOIST AND CLEAN.

SCRAPE THE TROWEL CLEAN AS YOU WORK. IF THE SCRAPINGS ARE CLEAN AND SOFT, MIX IT WITH THE STUFF ON THE HAWK. TRY TO KEEP THE COMPOUND TOGETHER SO IT WON'T DRY OUT AS FAST. IF THERE IS ANY PROBLEM WITH IT (HARD LUMPS, DIRT, ETC.), DUMP IT.

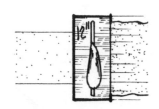 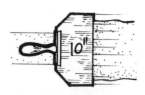  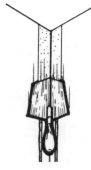

FOUR TROWELS ARE REQUIRED TO DO A GOOD JOINT JOB; A 6", 10", 12" AND A CORNER TROWEL.

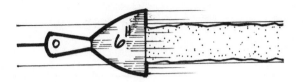

TO DO A JOINT, START WITH A 6" TROWEL AND PUT A LAYER OF COMPOUND AS WIDE AS THE TAPE THE LENGTH OF THE JOINT.

THE TROWEL HELD AT A FLATTENED ANGLE WILL LEAVE A NICE BED OF COMPOUND

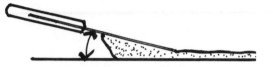

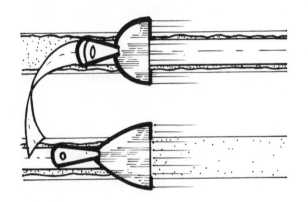

LAY THE TAPE ON THE COMPOUNDED JOINT AND PUSH IT FLAT WITH THE 6" TROWEL. GET ALL THE BUBBLES OUT.

SPREAD A THIN LAYER OF COMPOUND OVER THE TAPE. KEEP IT SMOOTH; LESS SANDING THAT WAY.

FIBERGLASS TAPE WITH A STICKY FACE ELIMINATES THE FIRST STEP: YOU JUST STICK IT ON AND THEN APPLY A COAT OF COMPOUND. IT'S A BIG HELP WHEN WORKING OVERHEAD, BUT ABOUT THREE TIMES THE COST OF PAPER TAPE.

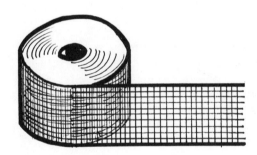

RAISE THE TROWEL TOWARD PERPENDICULAR FOR SMOOTHING. TRY DIFFERENT ANGLES FOR THE BEST RESULTS.

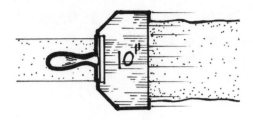

THE NEXT COAT IS DONE THE SAME WAY, BUT WITH THE 10" TROWEL. THAT IS, LAY ON A COAT...

WIPE THE TROWEL CLEAN...

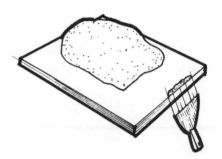

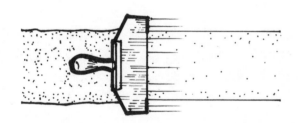

THEN RUN THE FULL LENGTH OF THE JOINT TO SMOOTH IT OUT.

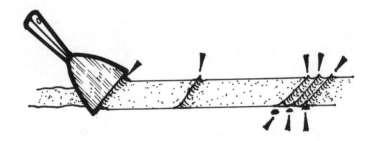

EVERY TIME THE TROWEL IS PICKED UP FROM THE JOINT IT PULLS SOME COMPOUND WITH IT LEAVING A RIDGE, SO TRY TO DO THE JOINT IN ONE STROKE. EVERY BUMP THE TROWEL HITS REFLECTS ON THE SURFACE OF THE JOINT. WHEN SMOOTHING OUT IT MAY TAKE A FEW STROKES TO GET THE EXCESS OFF. MAKE THAT LAST STROKE A NICE ONE, SINCE IT'S THE BASE FOR THE NEXT COAT.

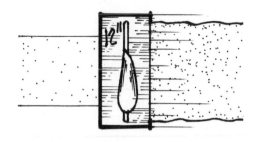

THE LAST COAT IS DONE WITH THE 12" TROWEL IN THE SAME MANNER AS THE OTHERS.

TRY TO MAKE A LONG SMOOTH RUN WITH THIS COAT.

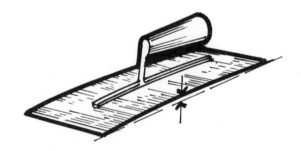

THE 12" TROWEL IS MUCH STIFFER THAN THE OTHERS AND IT HAS A CURVED BOTTOM, SO IT TAKES A GOOD BIT OF PRESSURE TO SMOOTH AND FEATHER OUT THE FINAL COAT.

EACH COAT MUST BE SANDED WHEN DRY WITH 80 GRIT. SAND THE LAST WITH 120 OR FINER. THE BETTER THE JOINTS ARE FEATHERED, THE LESS SANDING WILL BE REQUIRED.

A GOOD PLACE TO USE UP THE SEMI DRIED COMPOUND IS IN THE NAIL HOLES AND DENTS. RUN THE TROWEL AT A FLATTENED ANGLE TO LEAVE A LAYER OF COMPOUND IN THE DENT.

THEN HOLD THE TROWEL ALMOST PERPENDICULAR TO SCRAPE THE SURFACE CLEAN. GO TO THE NEXT DENT, DEPOSIT, SCRAPE, NEXT, DEPOSIT, SCRAPE, AND SO ON.

INSIDE CORNERS ARE DONE IN A SIMILAR FASHION. BUT HERE THE PAPER TAPE IS PRE-FOLDED TO FIT THE CORNER. THE TAPE HAS A CREASE DOWN THE MIDDLE SO FOLDING IS EASY. THERE ARE FOLDING TOOLS AVAILABLE.

THE PROCEEDURE IS THE SAME, BUT USE THE CORNER TROWEL TO LAY IN THE FIRST COAT OF COMPOUND....

LAY IN THE TAPE....

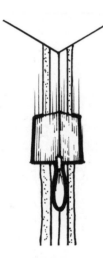

A COAT OF COMPOUND, SAND, AND SMOOTH ON A FINISH COAT OF COMPOUND.

EXTERIOR CORNERS ARE THE EASIEST OF ALL. NAIL ON THE CORNER BEAD AND COMPOUND THE JOINTS WITH THE 6" TROWEL.

THERE WILL BE EXCESS COMPOUND RUNNING AROUND THE CORNER AS YOU TROWEL, BUT NOT TO WORRY, IT'S EASY TO SCRAPE AND SAND IN THIS SPOT.

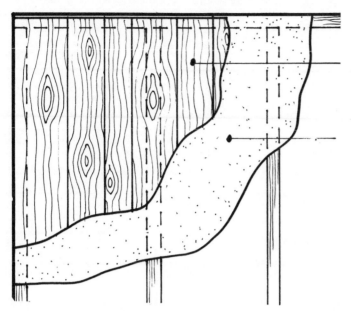

IF A WALL IS TO BE COVERED WITH PLY-WOOD PANEL SHEETS, THE BEST INSTAL-LATION IS OVER $3/8"$ SHEETROCK. IT'S SOLID AND MAKES FOR BETTER SOUND PROOFING.

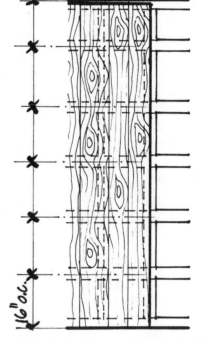

THE NEXT BEST WAY IS ON HORIZONTAL STRAPPING. (16" O.C.)

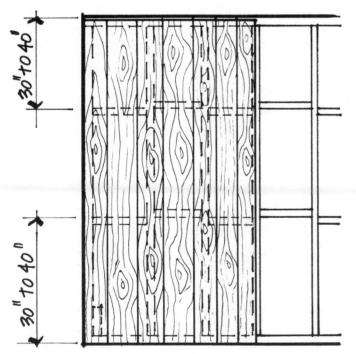

IF APPLIED DIRECTLY TO STUDS, BLOCK THE STUDS AT 30" TO 40" FROM TOP AND BOTTOM.

YAK
YAK
YAK...

THIS WILL PROTECT THE AREAS WHERE BUMPING OCCURS.

PANELING

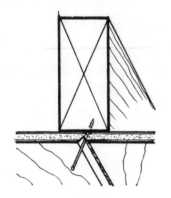

WHEN NAILING IN THE "V" GROOVE, NAIL THROUGH THE SIDE OF THE "V" AT A SLIGHT ANGLE; IT HOLDS BETTER.

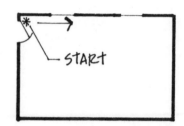

START

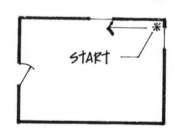

START

WITH ANY TRIM, LOOK FOR WAYS TO HIDE JOINTS. WITH PANELING, THE LAST PIECE IS THE TOUGHEST TO FIT, SO TRY TO FIND AN EASY WAY. A DOOR OR WINDOW NEAR A CORNER IS A GOOD PLACE TO MAKE THAT LAST JOINT EASY.

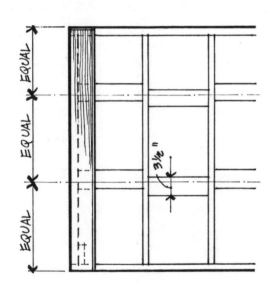

WITH SOLID ¾" PANELING, LAID UP VERTICALLY, 2×4 HORIZONTAL BLOCKING IS NEEDED. IF THE BLOCKS ARE PUT IN FLAT AND STAGGERED, THEY ARE EASIER TO NAIL AND IT LEAVES A STRAIGHT LINE OF WOOD TO NAIL INTO. YOU DON'T HAVE TO GUESS WHERE THE NAILING IS.

BEFORE ANY PANELING IS NAILED UP, IT IS A GOOD IDEA TO SPREAD IT AROUND THE ROOM SO IT CAN BE LOOKED OVER. MATCH THE PANELING, LOOK FOR VERY DARK OR LIGHT; PICK OUT ODD LOOK-ING ONES. CUT UP THOSE THAT DON'T BLEND IN AND USE FOR TRIM.

BACK PRIME IF THERE IS A MOISTURE PROBLEM.

IF THE PANELS ARE TO BE PAINTED OR STAINED DARK, IT IS A GOOD IDEA TO PAINT THE TONGUES SO THEY WON'T SHOW WHEN THE BOARDS SHRINK.

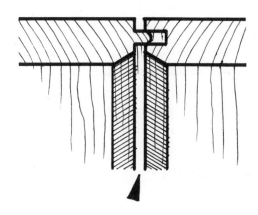

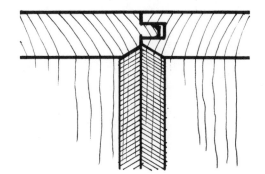

THE JOINTS DON'T ALWAYS STAY LIKE THIS.

USE A SCRAP PIECE OF PANELING (THE GROOVE EDGE) AS A BLOCK TO HAMMER AGAINST AND DRIVE EACH PANEL UP TIGHT.

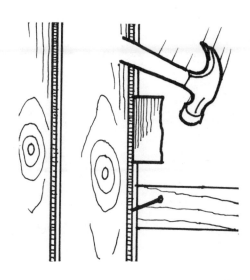

PANELING

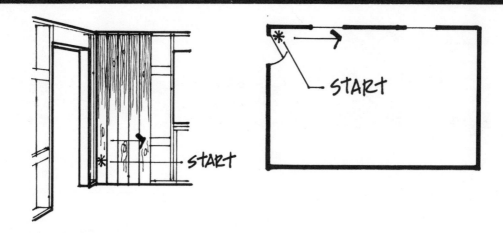

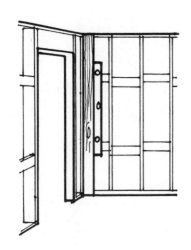

START IN A CORNER THAT WILL MAKE THE LAST PIECE EASY TO INSTALL.

PLUMB AND SCRIBE, IF NECESSARY, THE FIRST BOARD. TOE NAIL JUST ABOVE THE TONGUE AT THE BLOCKING AND SET THESE NAILS. FACE NAIL AT THE TOP AND BOTTOM WHERE THE BASE AND MOLDING WILL COVER NAILS.

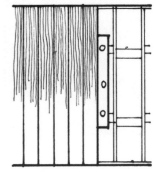

THERE IS NO GREAT MYSTERY TO NAILING EACH PANEL, BUT DO CHECK THE PLUMB ONCE IN A WHILE AND ADJUST.

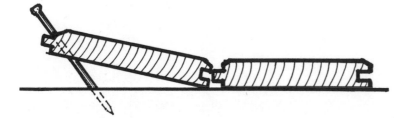

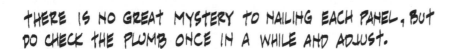

START A NAIL WITH THE TONGUE EDGE OF A PANEL RAISED SLIGHTLY TO AID IN DRIVING A STUBBORN BOARD OVER.

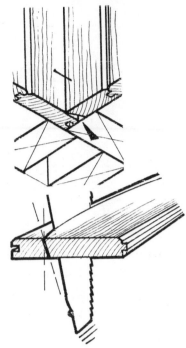

INSIDE CORNERS WILL HAVE TO BE SCRIBED WITH THE BOARD HELD IN THE PLUMB POSITION.

RIP WITH A HAND SAW AND BACK CUT FOR A TIGHT FIT.

THE OUTSIDE CORNER IS MITERED AND RELIEVED AT THE BACK OF THE MITER SO THAT THE FRONT OF THE MITER WILL BE TIGHT.

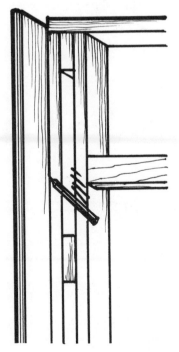

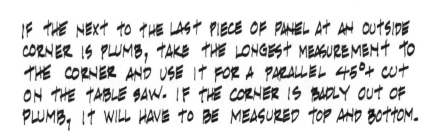

IF THE NEXT TO THE LAST PIECE OF PANEL AT AN OUTSIDE CORNER IS PLUMB, TAKE THE LONGEST MEASUREMENT TO THE CORNER AND USE IT FOR A PARALLEL 45°+ CUT ON THE TABLE SAW. IF THE CORNER IS BADLY OUT OF PLUMB, IT WILL HAVE TO BE MEASURED TOP AND BOTTOM.

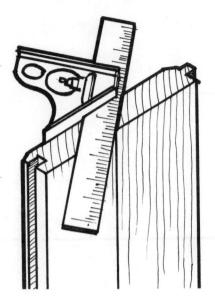

TRANSFER THE MARKS ON THE BACK, TO THE FRONT WITH A 45° COMBINATION SQUARE AND RIP WITH A HAND SAW.

A SHARP RIP SAW DOES THIS JOB NICELY.

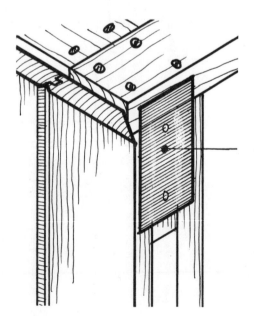

BUILDING PAPER FOR SHIMMING

A LITTLE FUDGING CAN BE DONE WITH BUILDING PAPER.

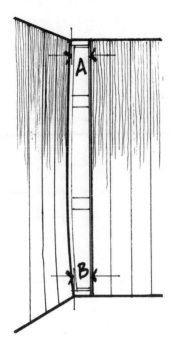

THE LAST SPACE IS MEASURED TOP AND BOTTOM.

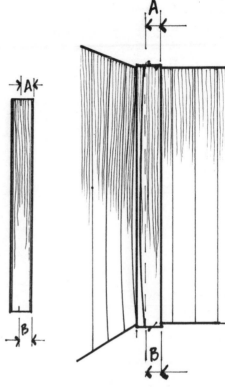

MARK THE PANEL, TACK IN PLACE AND SCRIBE.

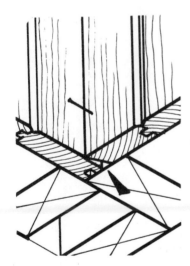

THIS SCRIBED LINE IS RIPPED WITH A SHARP RIP SAW, BEING SURE TO BACK-CUT IT.

HMMM...

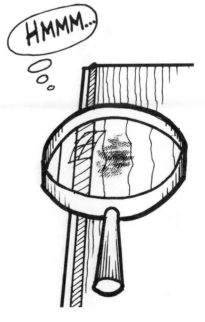

WHEN WORKING WITH PANELING, BE EXTRA CAREFUL ABOUT KEEPING HANDS CLEAN. IT'S EASIER THAN CLEANING THE PANELING.

DOOR FRAMES

MORE REAL FINISH WORK, DOOR FRAMES. THERE ARE TWO THINGS TO CONSIDER HERE, STYLE AND WIDTH.

FOR STYLE, THERE ARE TWO CHOICES; RABBETED AND PLANTED-ON STOP. I PREFER THE PLANTED STOP BECAUSE IT HELPS OVERCOME MANY DOOR PROBLEMS.

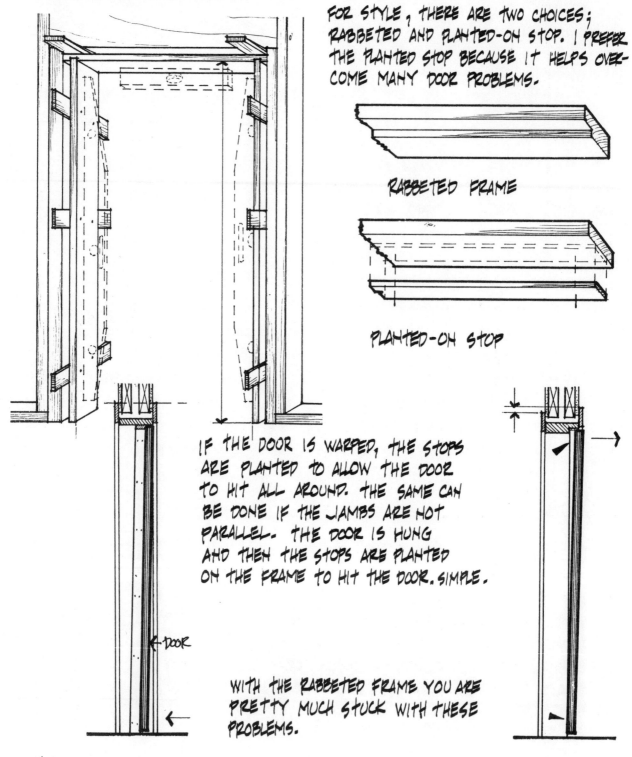

RABBETED FRAME

PLANTED-ON STOP

IF THE DOOR IS WARPED, THE STOPS ARE PLANTED TO ALLOW THE DOOR TO HIT ALL AROUND. THE SAME CAN BE DONE IF THE JAMBS ARE NOT PARALLEL. THE DOOR IS HUNG AND THEN THE STOPS ARE PLANTED ON THE FRAME TO HIT THE DOOR. SIMPLE.

←DOOR

WITH THE RABBETED FRAME YOU ARE PRETTY MUCH STUCK WITH THESE PROBLEMS.

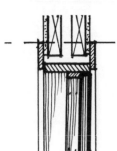

ANOTHER GOOD REASON FOR THE PLANTED-ON STOP IS THAT THE HEAD CASINGS WILL ALWAYS BE AT THE SAME LEVEL.

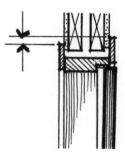

NOT SO WITH THE RABBETED FRAME:

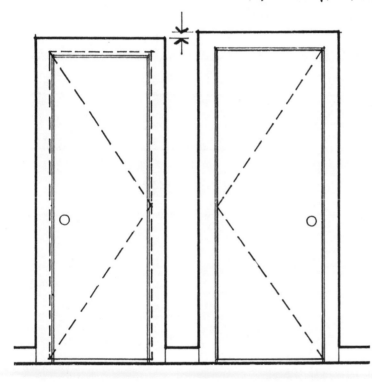

THE INSIDE IS HIGHER THAN THE OUTSIDE.

OUTSWING DOOR   INSWING DOOR

THE WIDTH IS DECIDED BY MEASURING FROM FACE OF SHEETROCK TO FACE OF SHEETROCK AT THE DOOR OPENINGS. YOU WILL BE SURPRISED AT THE VARIATIONS, NOT ONLY FROM DOOR TO DOOR, BUT ALSO TOP TO BOTTOM AND SIDE TO SIDE. OF COURSE, THE MORE CAREFULLY THE FRAMING WAS DONE, THE FEWER THE PROBLEMS LIKE THIS THAT WILL OCCUR. A GOOD FRAMER IS WORTH TWO FINISH MEN.

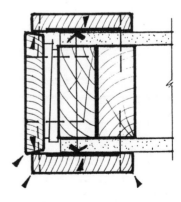

DOOR FRAMES

THE FRAMES SHOULD BE A TAD WIDER THAN THE MEASUREMENT DECIDED ON AND BEVELED BACK ON BOTH EDGES.

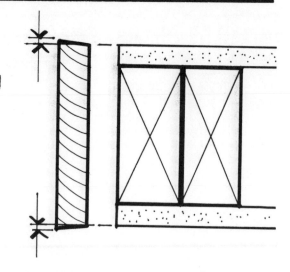

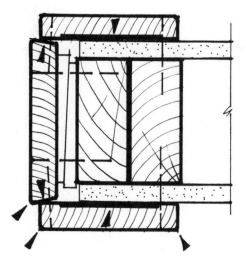

THIS WILL ALLOW FOR A GOOD JOINT BETWEEN THE CASING AND THE FRAME. A SHARP PLANER BLADE ON A TABLE SAW DOES A GOOD JOB. AS THE PIECES ARE BEVELED, MARK THE INSIDE FACE SO THAT THE RIGHT FACE WILL BE GROOVED FOR THE HEAD FRAME AND THE FRAME WILL BE ASSEMBLED PROPERLY.

THERE IS ALSO A CHOICE WHEN PUTTING FRAMES TOGETHER. EITHER GROOVE OUT THE JAMBS TO RE- CEIVE THE HEADER...

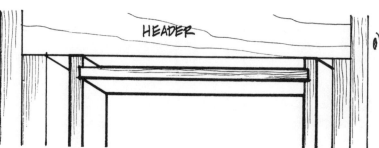

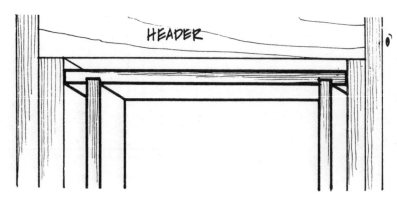

OR GROOVE OUT THE HEADER TO RECEIVE THE JAMBS. I PREFER TO GROOVE THE HEADER; IT'S MORE STABLE IN THE ROUGH OPENING.

42

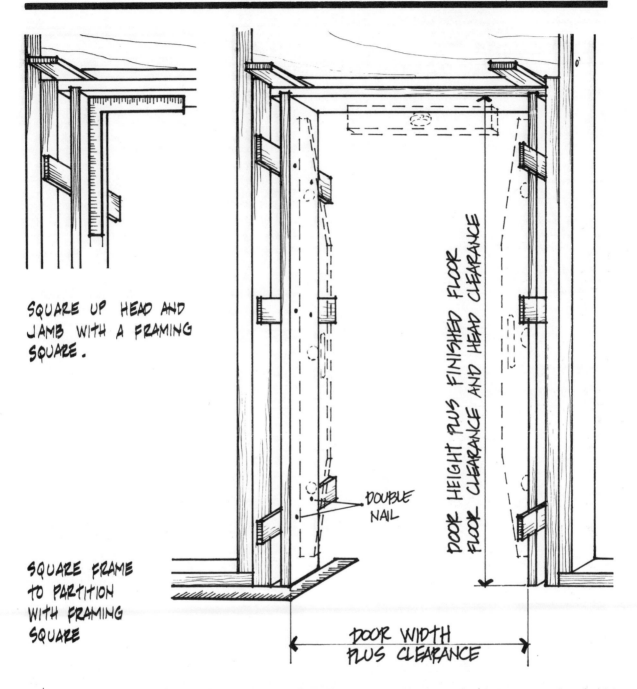

SQUARE UP HEAD AND
JAMB WITH A FRAMING
SQUARE.

SQUARE FRAME
TO PARTITION
WITH FRAMING
SQUARE

DOUBLE
NAIL

DOOR HEIGHT PLUS FINISHED FLOOR
FLOOR CLEARANCE AND HEAD CLEARANCE

DOOR WIDTH
PLUS CLEARANCE

IT'S A GOOD IDEA TO CUT ALL THE HEADERS AND JAMBS FOR THE FRAMES AT THE
SAME TIME. CUT THEM A LITTLE LONGER THAN REQUIRED AND TRIM TO FIT LATER.
THE JAMBS WILL ALL BE THE SAME LENGTH; THE DOOR HEIGHT PLUS THE DEPTH
OF THE GROOVE IN THE HEADER, HEADER CLEARANCE, THRESHOLD CLEARANCE AND
FINISHED FLOOR THICKNESS. THE HEADERS SHOULD BE AT LEAST 1" LONGER
THAN THE ROUGH OPENING WIDTH AND CUT TO FIT AFTER ASSEMBLY.

43

DOOR FRAMES

THE GROOVES IN THE HEADER ARE BEST CUT WITH A DADO BLADE ON THE TABLE SAW, BUT A KNIFE, HAND SAW AND CHISEL WILL DO A FINE JOB.

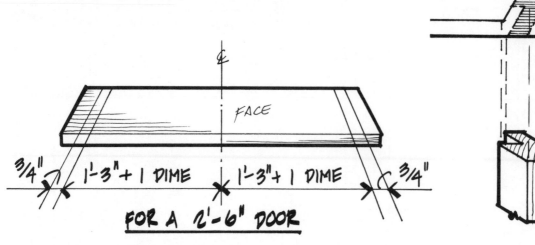

FACE

3/4"   1'-3" + 1 DIME   1'-3" + 1 DIME   3/4"

FOR A 2'-6" DOOR

START FROM THE CENTER OF THE HEADER AND MARK RIGHT AND LEFT 1/2 THE DOOR WIDTH PLUS THE CLEARANCE REQUIRED.

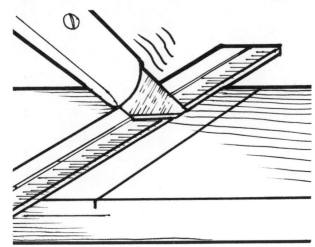

IF IT IS A HAND SAW AND CHISEL JOB, START BY SCORING, ALONG A COMBINATION SQUARE, WITH A SHARP KNIFE BEFORE CUTTING WITH A SHARP FINISH SAW. THIS WILL GIVE A NICE, CLEAN, POSITIVE LINE. CUT BOTH SIDES OF THE GROOVE ONLY AS DEEP AS REQUIRED (ABOUT 1/4") BEFORE CHISELING.

USE THE FINISHED HEADER AS A PATTERN FOR MAKING DUPLICATE HEADERS. LESS CHANCE FOR ERROR THAT WAY.

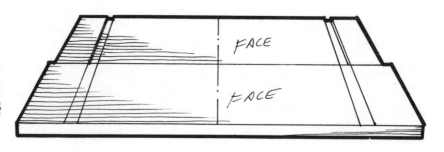

FACE

FACE

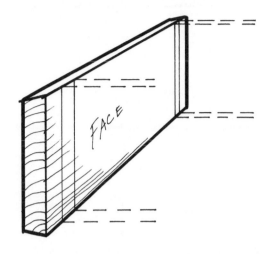

BEFORE GROOVING AND ASSEMBLING, CHECK FOR WHAT FACE GOES WHERE; IT'S EASY TO MESS UP HERE.

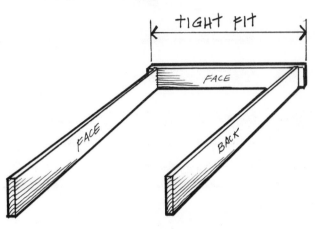

WITH THE FRAME ASSEMBLED, CUT THE HEADER FOR A TIGHT FIT IN THE ROUGH OPENING.

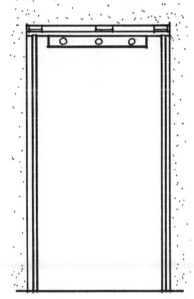

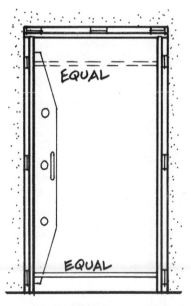

SLIP THE WHOLE BUSINESS INTO THE OPENING AND LEVEL THE HEADER BY SHIMMING THE BOTTOMS OF THE JAMBS. ONCE LEVELED, SNUG IT DOWN BY WEDGING SHINGLE TIPS DIRECTLY OVER THE JAMBS.

SHIM THE BOTTOM SIDE OF ONE JAMB UNTIL PLUMB USING A JAMB LEVEL (A GREAT TOOL FOR DOOR WORK). A REGULAR 6 FOOT LEVEL WILL WORK TOO. THEN SHIM THE MIDDLE TO A STRAIGHT LINE.

THE OPPOSITE JAMB CAN THEN BE SHIMMED OVER PARALLEL BY USE-ING A CUT-TO-SIZE MEASURING STICK.

DOOR FRAMES

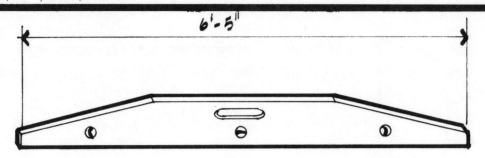

6'-5"

THIS DOOR JAMB LEVEL IS ALSO A STRAIGHT EDGE

WHEN SHIMMING WITH SHINGLES, USE TWO OPPOSING EACH OTHER EXCEPT WHEN THE FRAMING BEHIND IS TWISTED. IN SUCH CASES YOU MIGHT NEED TWO IN THE SAME DIRECTION TO MAKE THE TRIM PIECE COME SQUARE WITH THE WORK.

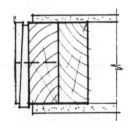

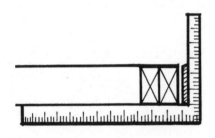

USE A FRAMING SQUARE AT THE BASE OF THE OPENING TO SQUARE THE JAMB FRAME WITH THE WALL.

CHECK THE HEADER FOR SQUARE WITH THE JAMB.

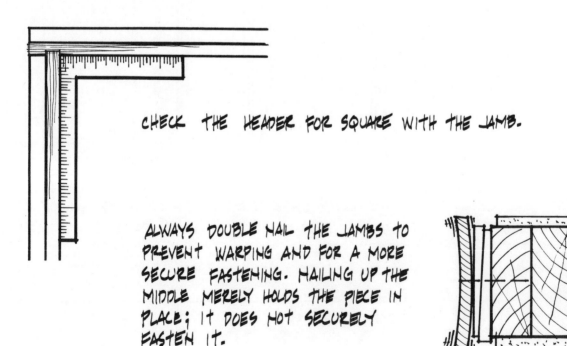

ALWAYS DOUBLE NAIL THE JAMBS TO PREVENT WARPING AND FOR A MORE SECURE FASTENING. NAILING UP THE MIDDLE MERELY HOLDS THE PIECE IN PLACE; IT DOES NOT SECURELY FASTEN IT.

IF THE DOORS HAVEN'T BEEN PRE-HUNG, NOW IS THE TIME TO HANG THEM. THERE SHOULD BE GOOD CLEARANCE ON EACH SIDE, TOP AND BOTTOM.

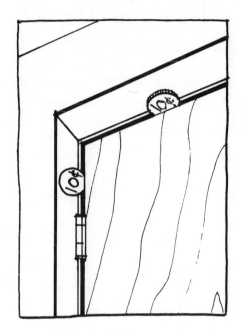

ABOUT ONE DIME'S THICKNESS AT THE TOP AND SIDES IF THE DOOR IS TO BE STAINED.

ABOUT ONE NICKLE'S THICKNESS AT THE TOP AND SIDES IF THE DOOR IS TO BE PAINTED.

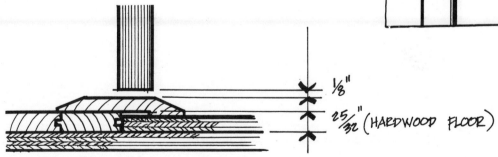

1/8"

25/32" (HARDWOOD FLOOR)

1/8" PLUS THE FINISHED FLOOR AND THRESHOLD IF ONE IS TO BE USED.

47

DOORS

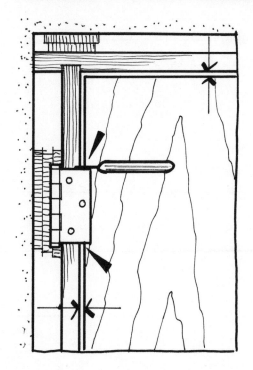

IF ALL IS SQUARE AND PLUMB, THE DOOR SHOULD FIT WITH NO TROUBLE. SOME TEMPORARY STOP PIECES WILL HOLD THE DOOR IN PLACE WHILE MARKING THE BUTT LOCATIONS. SHIM THE DOOR TO THE PROPER HEIGHT AND MARK THE JAMB AND DOOR WITH A SHARP KNIFE.

THE DOOR BUTTS (HINGES) ARE USUALLY SET AT 5" AND 10". I ALWAYS THINK OF WOOLWORTH'S 5 & 10 STORE AS A REMINDER.

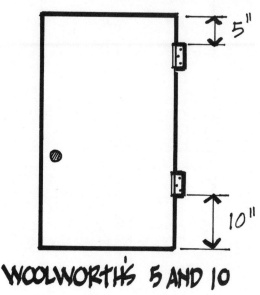

5"

10"

WOOLWORTH'S 5 AND 10

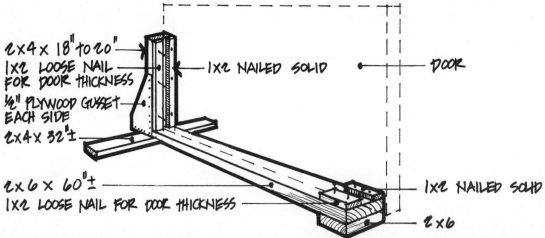

2x4 x 18" to 20"

1x2 LOOSE NAIL FOR DOOR THICKNESS

½" PLYWOOD GUSSET EACH SIDE

2x4x 32"±

1x2 NAILED SOLID

DOOR

2x6 x 60"±

1x2 LOOSE NAIL FOR DOOR THICKNESS

1x2 NAILED SOLID

2x6

A DOOR JACK IS A GREAT HELP WHEN WORKING ON DOORS. SOME CARPENTERS USE TWO 2x4 x 32" FEET, BUT I FIND THE SECOND ONE A FOOT TRIPPER. THE JACK SIMPLY SUPPORTS THE DOOR IN A SOLID VERTICAL POSITION.

48

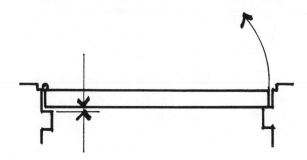

HINGE BINDING IS A BIG PROBLEM WITH DOORS, BUT CAN EASILY BE AVOIDED IF ENOUGH CLEARANCE IS LEFT BETWEEN THE DOOR STOP AND THE DOOR.

IF THE DOOR IS HINGE BOUND, THE HINGE MUST BE MOVED AWAY FROM THE STOP OR THE STOP MOVED FROM THE DOOR.

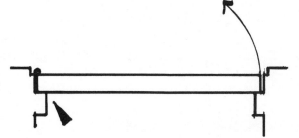

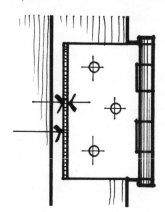

IF THE HINGE IS MOVED BACK, A WOOD SHIM SHOULD BE LAID IN TO FILL THE GAP AND KEEP THE HINGE IN PLACE.

IF A DOOR IS NOT PLUMB IT WILL EITHER WANT TO OPEN BY ITSELF OR STAY CLOSED.

OPEN SESAME

LATCH EDGE

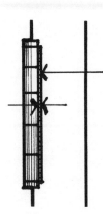

CARDBOARD SHIM

CARDBOARD SHIMS BEHIND THE HINGE WILL CHANGE THE HANG OF THE DOOR. SHIM THE TOP HINGE TO LOWER THE LATCH EDGE. SHIM THE BOTTOM TO RAISE THE LATCH EDGE.

49

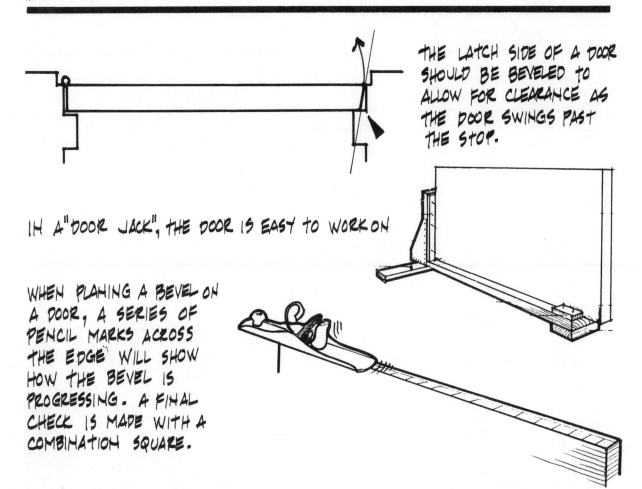

THE LATCH SIDE OF A DOOR SHOULD BE BEVELED TO ALLOW FOR CLEARANCE AS THE DOOR SWINGS PAST THE STOP.

IN A "DOOR JACK", THE DOOR IS EASY TO WORK ON

WHEN PLANING A BEVEL ON A DOOR, A SERIES OF PENCIL MARKS ACROSS THE EDGE WILL SHOW HOW THE BEVEL IS PROGRESSING. A FINAL CHECK IS MADE WITH A COMBINATION SQUARE.

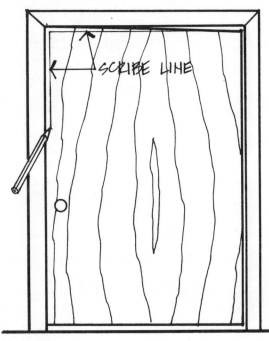

SCRIBE LINE

IF THE DOOR HAS PROPER CLEARANCE ON THE HINGE SIDE AND HITS THE HEADER OR THE OPPOSITE JAMB, THE DOOR MUST BE PLANED TO FIT. CLOSE THE DOOR AS FAR AS IT WILL GO AND SCRIBE WITH A PENCIL, ALLOWING FOR CLEARANCE. PULL THE HINGE PINS, PUT THE DOOR IN THE DOOR JACK AND PLANE IT DOWN.

FOR SETTING LOCKS, BORROW OR RENT A LOCK BORING KIT THAT IS MADE FOR THE LOCKSET BEING INSTALLED. LUMBER YARDS HAVE THESE. OTHERWISE, IT'S MEASURE AND DRILL WITH A BRACE AND BIT.

I HAVE PUT A LOT OF HINGES ON WITH A COMBINATION SQUARE, A SHARP KNIFE, AND A SHARP CHISEL.

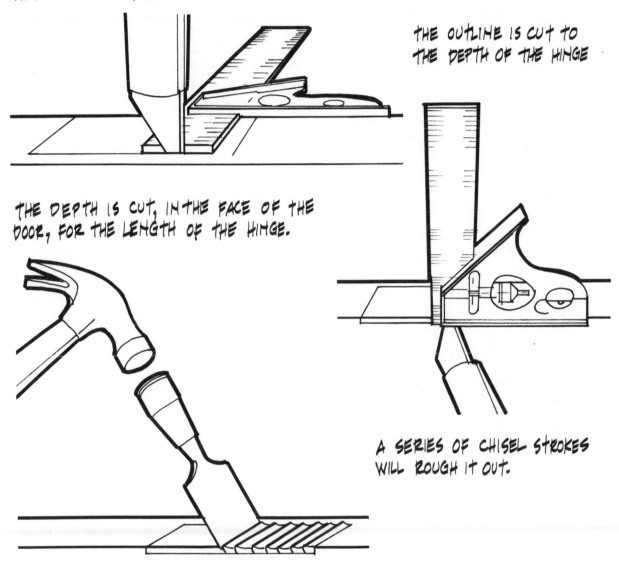

THE OUTLINE IS CUT TO THE DEPTH OF THE HINGE

THE DEPTH IS CUT, IN THE FACE OF THE DOOR, FOR THE LENGTH OF THE HINGE.

A SERIES OF CHISEL STROKES WILL ROUGH IT OUT.

WITH THE CHISEL HELD FLAT, THE JOB IS FINISHED OFF. WHEN I WORK A CHISEL WITH MY HANDS, I WORK WITH RESTRAINT. ONE HAND GUIDES WHILE THE OTHER PUSHES. A SHARP TOOL CAN BE CONTROLLED EASIER THAN A DULL ONE.

51

STANLEY MAKES A BUTT MARKING TOOL THAT COMBINES ALL THREE STEPS IN ONE TOOL. ONE ARM SETS FOR THE WIDTH, THE OTHER FOR THE DEPTH. THIS SCRIBING TOOL HAS SHARP EDGES FOR CUTTING THE OUTLINE, LIKE THE KNIFE AND SQUARE SET UP. THE CHISEL WORK IS STILL REQUIRED

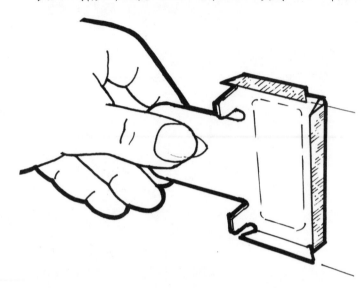

STANLEY ALSO MAKES A BUTT MARKER THAT IS PUT IN PLACE AND HIT WITH A HAMMER. THE SHARP EDGES LEAVE THE OUT- LINE READY FOR CHISELING. IT COMES IN MANY SIZES.

ALL THESE TOOLS ARE GOOD, BUT THE VERY BEST SYSTEM FOR DOING MANY DOORS IS THE ROUTER AND BUTT TEMPLATE GUIDE. ONCE SET UP, JAMBS AND DOORS ARE ROUTED QUICKLY AND ACCURATELY. THE GUIDE IS EXPENSIVE AND CANNOT BE RENTED.

BUTTS ARE SET WITH THE EDGE ABOUT ¼" FROM THE FACE OF THE DOOR AND THE SCREWS ARE OFFSET IN THE HOLES TO FORCE THE BUTT AGAINST THIS EDGE.

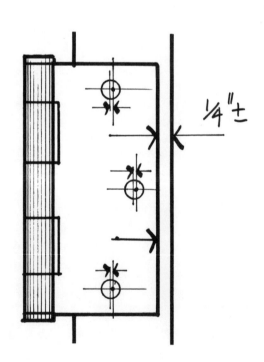

¼"±

DOOR CASING

NOW FOR SOME EXCITING FINISH WORK;
CASINGS. MARK THE REVEAL ON THE
EDGE OF THE FRAMES WITH A COMBINA-
TION SQUARE AS A GUIDE. BETTER STILL,
MAKE UP A MARKING GAUGE FROM A
PIECE OF 3/4" PINE. THE COMBINATION
SQUARE IS APT TO SLIP, MAKING SOME
FUNNY LOOKING REVEALS.

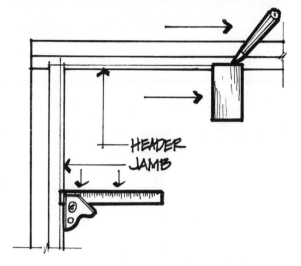

HEADER
JAMB

WINDOW OR DOOR FRAME

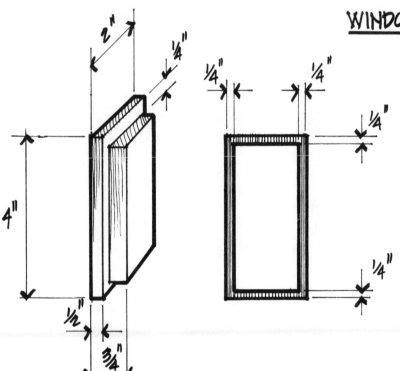

THE WOOD GAUGE IS ALWAYS
THE SAME AND IT FREES
THE COMBINATION SQUARE
FOR OTHER JOBS. 1/4" IS
ABOUT THE USUAL REVEAL,
BUT SUIT YOURSELF.

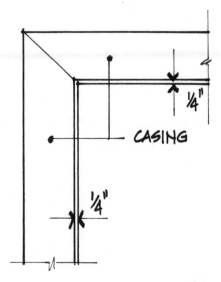

CASING

DOOR CASING

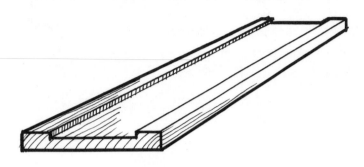

THERE ARE BASICALLY TWO CHOICES WITH CASINGS, MITERED AND SQUARE CORNERS, AND THEY ARE MORE EASILY FITTED IF THE BACKS ARE RELIEVED.

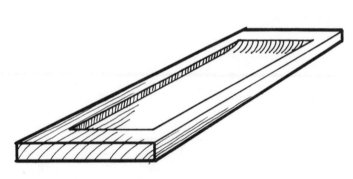

THIS IS PARTICULARLY HELPFUL ON BAD WALLS. IF SQUARE CORNER CASINGS ARE USED, THEY CANNOT BE RELIEVED TO THE ENDS BECAUSE THEY SHOW.

THE EASIEST WAY TO RELIEVE THE CASINGS IS WITH A DADO BLADE ON A TABLE SAW. THE MITERED CASINGS ARE RUN RIGHT THROUGH, BUT THE SQUARE ONES ARE STARTED SHORT AND ENDED SHORT.

THE BEST TOOL FOR TRIM WORK IS AN ELECTRIC MITER BOX. YOU CAN'T BEAT IT FOR SPEED AND ACCURACY. IF THIN SHAVING IS REQUIRED, THERE IS NO BETTER WAY THAN WITH THIS TOOL.

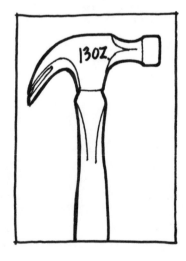

MOST CARPENTERS CARRY ONE HAMMER, A 16 OZ., AND IT WILL DO FOR ANY AND ALL JOBS, EVEN FINE FINISH WORK. I PREFER A 13 OZ. FOR FINISH WORK; THE CONTROL IS SO MUCH BETTER. THESE HAMMERS ARE NOT SO EASY TO FIND THESE DAYS, BUT I KNOW WOODCRAFT SUPPLIES IN MASSACHUSETTS HAS A NICE ONE.

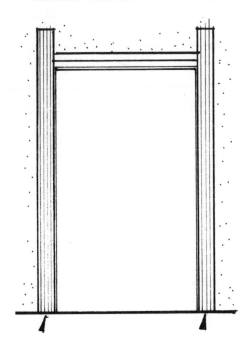

IF THE CASINGS GO ON BEFORE THE FLOOR, SLIP A LOOSE PIECE OF FLOORING UNDER THE CASING BEFORE ANY MEASUREMENTS ARE TAKEN. SQUARE THE BOTTOM OF THE SIDE CASINGS BY HOLDING THEM AGAINST THE REVEAL LINE AND SCRIBING TO THE FLOOR.

START ON ONE SIDE CASING AND MARK THE HEIGHT WITH A SHARP KNIFE AT THE REVEAL LINE ON THE HEAD FRAME.

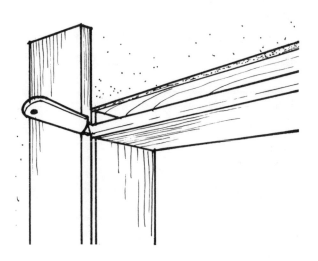

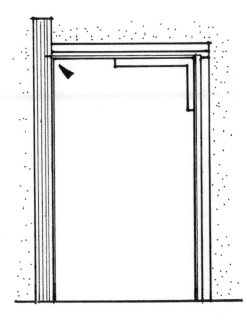

IF THE HEAD AND JAMB ARE SQUARE THEN A SQUARE CUT AT THE TOP OF THE SIDE CASING WILL WORK. IF THEY ARE NOT SQUARE, THEN A TRIAL CUT, A LITTLE LONGER THAN THE REVEAL MARK IS MADE AND A TRIAL FIT WITH THE HEAD CASING FOLLOWS. WHEN THE TRIAL FIT IS RIGHT, REPEAT IT AT THE PROPER LENGTH ON THE SIDE CASING. TACK IT IN PLACE.

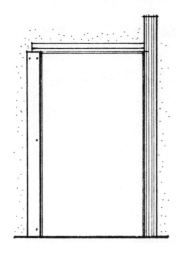

REPEAT FOR THE OTHER SIDE CASING.

SQUARE CUT ONE END OF THE HEAD CASING, HOLD IN PLACE AND MARK WITH A SHARP KNIFE. CUT AND HOLD IN PLACE TO CHECK THE FIT.

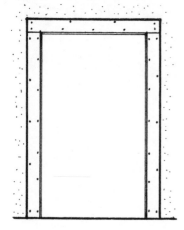

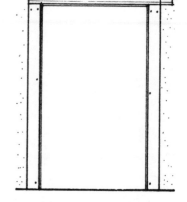

THE FINISHED PRODUCT.

BEFORE NAILING IN PLACE, I LIKE TO RUN A BEAD OF GLUE ON THE BACK OF THE CASING WHERE IT TOUCHES THE FRAME AND WHERE THE HEADER SITS ON THE SIDE CASINGS. BE CARE- FUL IF THE WOOD IS TO BE STAINED, BECAUSE STAIN WILL NOT TAKE ON A GLUED SURFACE.

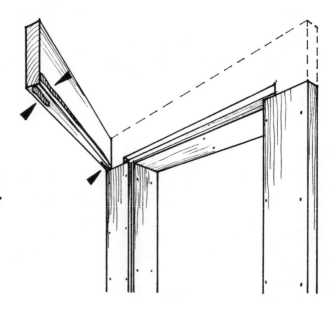

YOU CAN, OF COURSE, CUT ALL THE PIECES TO THE RIGHT LENGTHS, NAIL, SAND AND FILL THE GAPS AND IT MIGHT LOOK ALL RIGHT, BUT IF THE TRIM IS TO BE STAINED IT NEVER WILL LOOK GOOD UNLESS YOU CUSTOM FIT EACH CASING.

IF THERE IS BAD FRAMING AT THE OPENING, IT CAN BE OVERCOME BY SHIMMING OR SHAVING THE BACK OF THE CASING. HERE IS WHERE THE RELIEVING ON THE BACK-SIDE OF THE CASINGS HELPS.

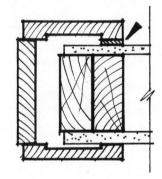

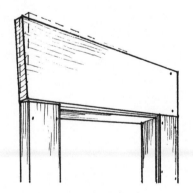

IF THE HEAD CASING TIPS BACK...

A WEDGE SHAPED SHIM WILL MAKE IT RIGHT. IT IS ALWAYS BEST WHEN THE FACE OF THE CASINGS ARE ON THE SAME PLANE.

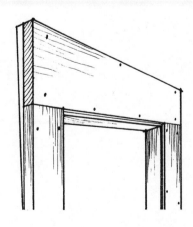

DOOR CASING

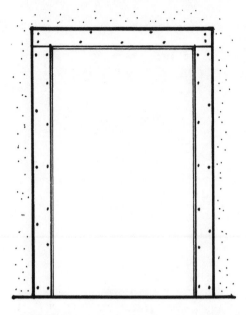

USE 8d FINISH OR CASING NAILS AT THE OUTER EDGE AND 4d OR 6d AT THE INNER EDGE. USE AS MANY NAILS AS IS REQUIRED TO SECURELY FASTEN THE TRIM IN PLACE.

CLACK CLACK

RAP THE TRIM WITH THE KNUCKLES AND LISTEN FOR THE TELL-TALE RATTLE OF LOOSENESS. ALSO NAIL WHEREVER THE TRIM IS PULLING AWAY FROM THE WALL OR FRAME.

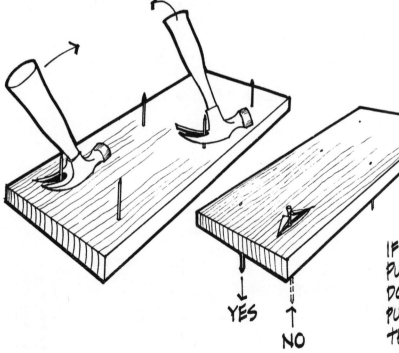

YES   NO

IF A PIECE OF TRIM MUST BE PULLED AND USED AGAIN, DON'T DRIVE THE NAILS OUT. PULL THEM THROUGH TO KEEP THE FACE CLEAN.

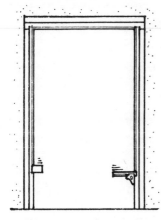

THE MITERED CASING IS STARTED THE SAME WAY AS THE SQUARE CASING, BY MARKING THE REVEAL LINE ON THE FRAME.

THE SIDE CASING SITS ON THE FINISHED FLOOR OR BLOCKS SIMULATING THE FINISHED FLOOR AND IS SCRIBED TO FIT.

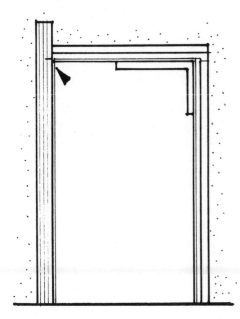

CHECK THE HEADER FOR SQUARE WITH THE JAMB AND MARK THE HEADER REVEAL LINE ON ONE SIDE CASING.

IF ALL IS SQUARE, THEN A 45° CUT IS MADE AND THE CASING IS TACKED IN PLACE.

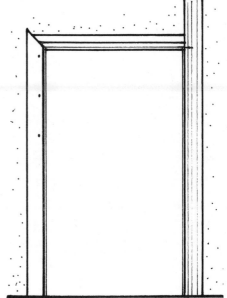

DOOR CASING

IF ALL IS NOT SQUARE, THEN A TRIAL CUT WITH SOME SCRAP PIECES WILL GIVE THE RIGHT FIT.

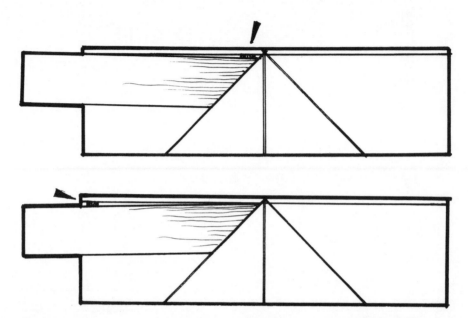

PUT WEDGES OF WHATEVER THICKNESS, AT THE RIGHT OR LEFT END OF THE PIECE OF TRIM IN THE MITER BOX TO GET THE DESIRED ANGLE. THEN DUPLICATE THAT CUT ON THE FIRST SIDE CASING AND TACK IT IN PLACE. DO THE SAME FOR THE OTHER SIDE CASING, BUT DON'T TACK IT IN PLACE UNTIL THE HEAD CASING LENGTH IS MARKED.

CUT A 45° OR THE CORRECTED ANGLE ON A PIECE OF HEADER STOCK. HOLD IN PLACE AND IF IT NEEDS CORRECTING SHAVE IT WITH THE MITER BOX OR A SHARP BLOCK PLANE. HOLD THE TRIM PIECE SECURELY WHILE PLANING; YOU MAKE CLEANER CUTS THAT WAY. WHEN THE FIT IS GOOD, MARK FOR LENGTH AT THE SIDE FRAME REVEAL LINE.

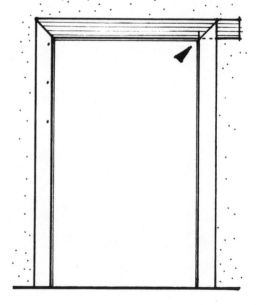

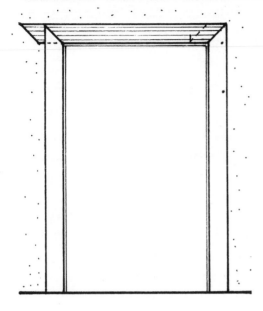

MAKE A TRIAL CUT PAST THE MARK AND SEE HOW IT FITS. WHEN THE FIT IS GOOD, DUPLICATE THE CUT AT THE CORRECT LENGTH

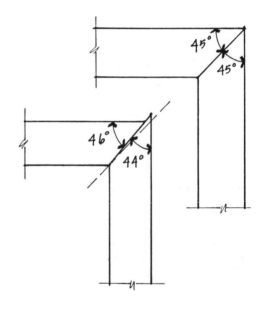

A MITERED JOINT MUST BE MADE UP OF PIECES CUT ON EQUAL ANGLES OR YOU GET AN ODD CORNER. YOU CAN FUDGE A LITTLE, ESPECIALLY IF THE TRIM IS TO BE PAINTED.

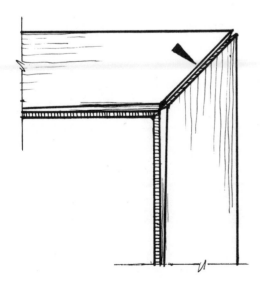

IF THE HEADER IS TIPPED BACK, THE MITER WILL BE OPEN AT THE FACE.

DOOR CASING

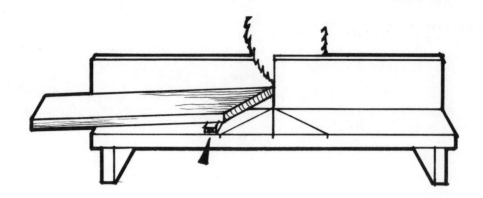

IF IT IS NOT TOO BAD IT CAN BE CORRECTED BY RAISING THE PIECE IN THE MITER BOX AT THE SAW END. THIS WILL RELIEVE THE BACK EDGE OF THE MITER. SHIMMING THE BACKSIDE WILL DO THE SAME THING AND IT WILL NOT SHOW IF PAINTED. GLUE IT THE SAME AS THE SQUARE CASING, INCLUDING THE MITER.

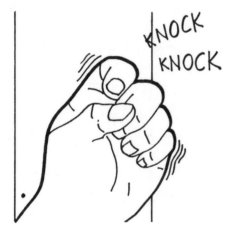

KNOCK
KNOCK

NAIL THE TRIM TO BE SECURE AND TEST BY KNOCKING.

THE TELL-TALE
CLACK CLACK
NEEDS NAILING.

CLACK
CLACK

IF THE TRIM IS TO BE PAINTED, SAND THE JOINTS WITH A PIECE OF SANDPAPER WRAPPED AROUND A BLOCK OF WOOD WHILE THE GLUE IS STILL WET. THE DUST WILL FILL IN NICELY, MAKING AN INVISIBLE JOINT. STAINED WORK MUST FIT WITH-OUT SANDING, BECAUSE EVERY SCRATCH WILL SHOW AFTER STAINING.

WINDOW CASING IS BASICALLY SIMILAR TO DOOR CASING. THE WINDOW FRAME HAS TO BE PACKED OUT OR PLANED BACK TO THE INTERIOR WALL FACE.

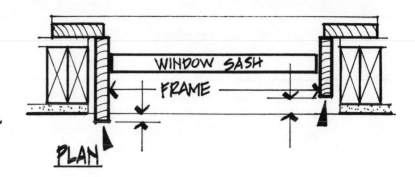

WINDOW SASH

FRAME

PLAN

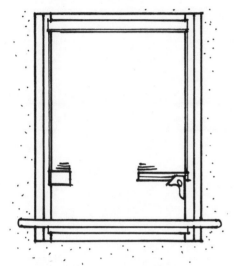

THEN IF THERE IS A REVEAL, IT IS MARKED ON THE FRAME.

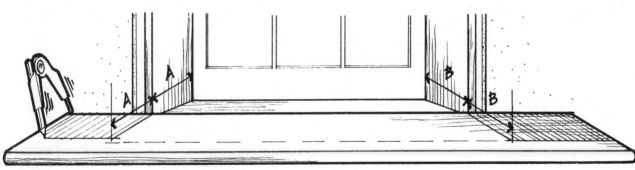

THE STOOL IS THEN SCRIBED TO FIT AGAINST THE WALL AND WINDOW, WITH A LITTLE BIT OF GAP AT THE WINDOW TO ALLOW FOR PAINT AND WINDOW CLEARANCE. (1/32" MINIMUM).

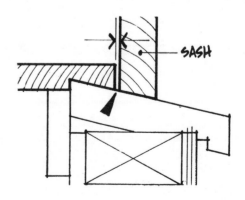

SASH

63

WINDOW CASING

THE STOOL LENGTH WILL BE ABOUT 3/4" PAST THE SIDE CASINGS, SO KEEP THE ROUGH PIECE PLENTY LONG. THE STOOL PIECE MIGHT HAVE TO BE RIPPED TO LEAVE NO MORE THAN 3/4" PAST THE FACE OF THE CASING AND APRON.

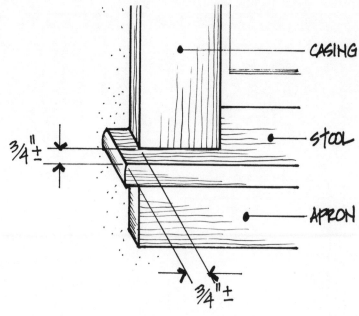

CASING

STOOL

APRON

3/4"±

3/4"±

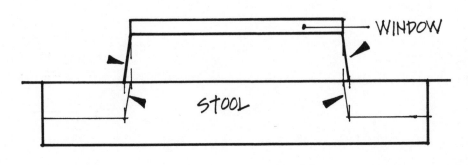

WINDOW

STOOL

THE CUT PARALLEL TO THE JAMB IS MARKED.

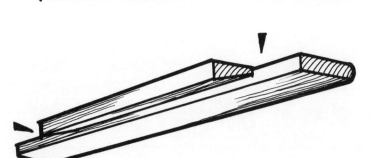

CUT THE TWO NOTCHES AND TACK THE STOOL IN PLACE.

NOW THE SIDE CASINGS ARE CUT ON THE BOTTOMS TO FIT THE STOOL. THEY SHOULD BE SQUARE CUTS. THE TOP CUTS ARE MARKED...

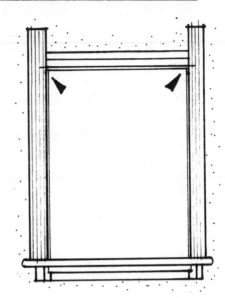

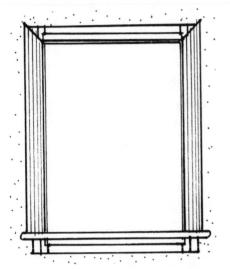

AND CUT.

ONE SIDE CASING IS TACKED IN PLACE AND A TRIAL HEADER END IS CUT AND FITTED. THE OPPOSITE END IS MARKED FOR LENGTH.

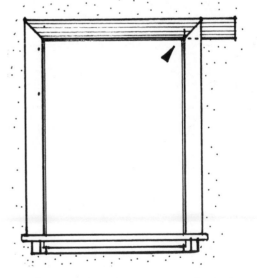

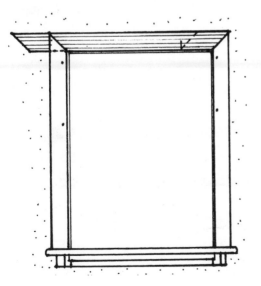

THE OTHER END IS TRIAL CUT AND FITTED.

65

WINDOW CASING

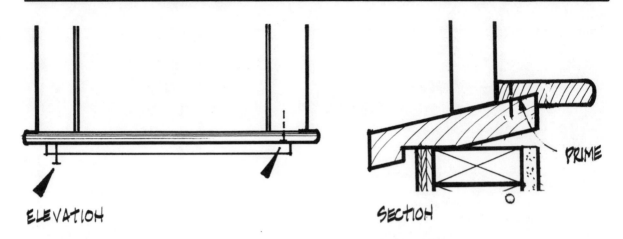

ELEVATION

SECTION

PRIME

THE STOOL IS NAILED IN PLACE, THEN THE CASINGS. IT'S A GOOD IDEA TO PRIME THE UNDERSIDE OF THE STOOL THAT SITS ON THE WINDOW SILL. DRIVE A NAIL FROM THE BOTTOM OF THE STOOL UP INTO THE SIDE CASINGS.

THE APRON REACHES FROM OUTSIDE OF CASING TO OUTSIDE OF CASING.

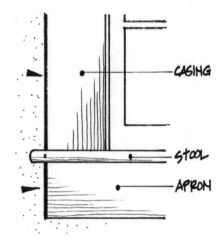

CASING

STOOL

APRON

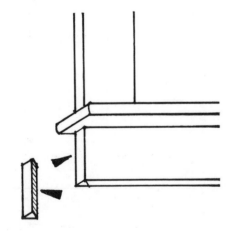

IF TRIM IS STAINED, CUT A 45° RETURN AT EACH END OF THE APRON. END GRAIN STAINS DARKER THAN FACE GRAIN.

66

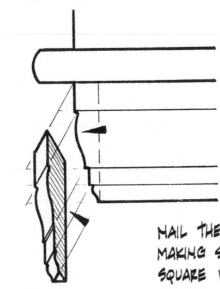

IF THE TRIM IS MOLDED, SHAPE THE RETURN WITH A COPING SAW WHEN TRIM IS TO BE PAINTED. OTHERWISE A 45° RETURN PIECE SHOULD BE FITTED.

NAIL THE APRON IN PLACE, MAKING SURE THE STOOL IS SQUARE WITH THE WINDOW.

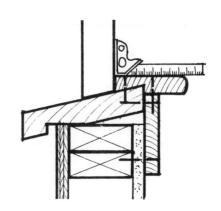

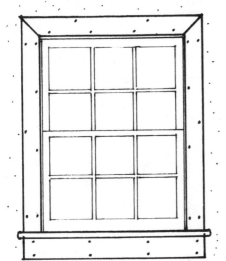

THE FINISHED PRODUCT.

A HEAT PIECE OF EQUIPMENT TO USE FOR CASING WORK IS THE "CRICKET." ITS LIGHT, THE RIGHT HEIGHT, STABLE, AND IT HAS A SHELF FOR CARRYING TOOLS AND THINGS AROUND ON. MORE ON THIS TIDBIT LATER.

BASEBOARD

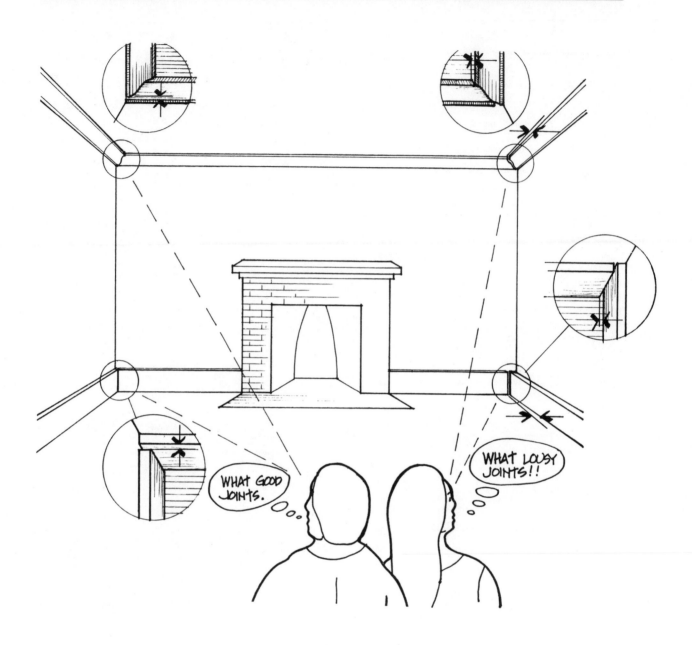

WHEN PUTTING IN BASEBOARD OR CEILING TRIM, KEEP IN MIND WHAT YOU SEE WHEN ENTERING A ROOM AND MAKE THE JOINTS IN SUCH A WAY THAT THEY LOOK GOOD NO MATTER HOW BAD THEY MIGHT BE. IN THIS CASE THE JOINTS ON BOTH SIDES OF THE ROOM ARE BAD, BUT THE JOINTS ON THE RIGHT ARE OBVIOUS AND THE ONES ON THE LEFT ARE NOT.

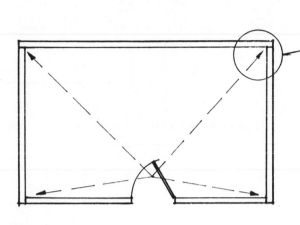

IF THE PIECE ON THE WALL OPPOSITE THE DOOR GOES IN FIRST, THE SEAM OF THE JOINT WILL BE TO THE SIDE AND HARD TO SEE.

PLAN

THE BEST JOINT IS A COPED JOINT. THERE IS A BUILDUP OF JOINT COMPOUND IN THE CORNERS SO THEY ARE RARELY SQUARE. THE COPED CORNER, WHEN SNAPPED IN PLACE, IS A VERY TIGHT FIT.

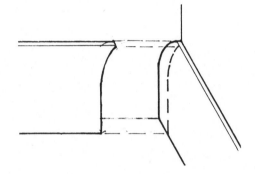

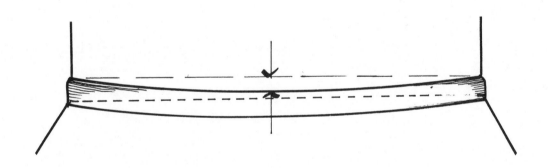

Oops, a little snug.

WITH THE COPED CORNER, THE FIRST PIECE IN IS CUT SQUARE ON EACH END AND A TAD LONG. THE PIECE IS BOWED AWAY FROM THE WALL AT THE CENTER AND PUSHED INTO PLACE AT THE ENDS. IF THE PIECE IS THE RIGHT LENGTH, IT WILL GENTLY SNAP FROM YOUR HAND WHEN THE CENTER IS MOVED CLOSER TO THE WALL. IF IT IS TOO LONG, IT JUST WON'T GO AT ALL. AND IF IT IS A LITTLE (AS OPPOSED TO A TAD) TOO LONG, THE CORNER WILL CRACK.

69

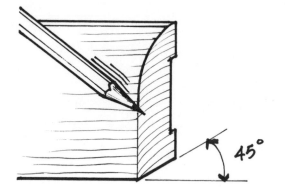

THE NEXT PIECE IS CUT AT A 45° ANGLE AS IF IT WERE A MITERED CORNER. RUB A PENCIL ON THE FRONT CORNER OF THIS MITER SO THAT IT WILL BE EASIER TO FOLLOW WITH THE COPING SAW.(HENCE THE "COPED CORNER").

START THE COPING SAW ON THE TOP, AT 90° OR PERPENDICULAR TO THE BACK AND, AS YOU FOLLOW THE PENCILED CORNER, EASE THE SAW BACK FROM PERPENDICULAR TO ABOUT 87°. OR START ON THE BOTTOM AT 87° AND WORK TOWARD THE TOP EASING OVER TO 90°.

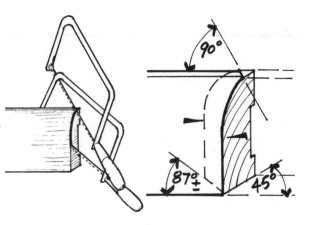

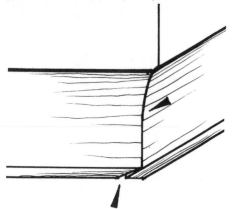

THIS WILL INSURE THAT THE FRONT EDGE WILL HIT HARD AGAINST THE ADJACENT PIECE OF BASE BOARD. MEASURE FROM THE FACE OF THE BASE BOARD ALREADY IN PLACE TO THE OPPOSITE WALL AND MARK THIS ON THE COPED PIECE. ADD A LITTLE TO MAKE A SNAP FIT. IT'S REALLY SIMPLE, BUT IT MIGHT TAKE A FEW PRACTICE CUTS TO GET THE HANG OF IT.

EVEN INTRICATE MOLDING SHAPES CAN BE CUT THIS WAY. CEILING MOLDING IS DONE THE SAME WAY.

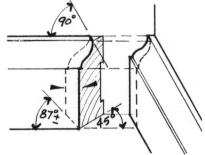

THE BEST LOOKING
CORNER FOR SQUARE
BASE IS THE COMBINA-
TION MITERED AND
COPED JOINT. THIS
SQUARE BASE USUALLY
HAS A ROUNDED UPPER
CORNER.

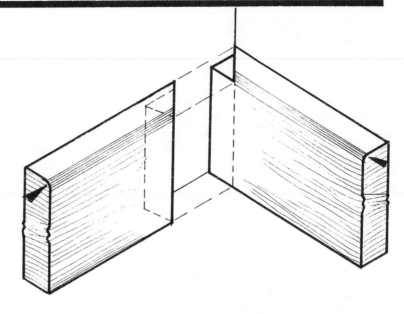

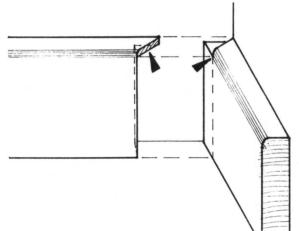

THE FIRST PIECE IS SQUARE CUT
AND THEN MITERED FOR ABOUT
1/8" AT THE TOP (TO THE BOTTOM OF
THE ROUND CORNER). THE ADJACENT
PIECE IS MITERED AND THEN CUT,
FROM THE BOTTOM, UP TO THE BOTTOM
OF THE ROUND CORNER. BACK CUT
FOR A TIGHT FIT IN FRONT. CUT
ACROSS WITH A KNIFE AND THERE
REMAINS ABOUT A 1/8" MITER LIP.

BASE BOARD IS NAILED AT THE STUDS
TOP AND BOTTOM. SOMETIMES THE BASE
IS NOT TIGHT AGAINST THE WALL BE-
TWEEN STUDS, BUT WE SHALL OVERCOME.
USE 16d FINISH NAILS HIGH ON THE
BASE AND ANGLE DOWN TO REACH THE
2X4 PLATE. WHEN THE NAIL IS SET
IT SHOULD PULL THE BASE IN.

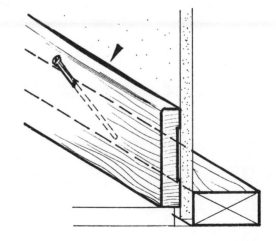

BASEBOARD

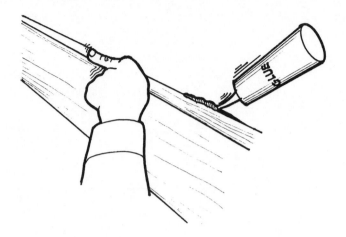

IF THERE IS STILL A SPACE BETWEEN BASE AND WALL, SQUEEZE SOME GLUE IN AND SMOOTH WITH A FINGER. IT MIGHT TAKE A FEW COATS. DON'T USE THIS METHOD IF THE TRIM IS TO BE STAINED.

WHEN "A" "DUCK PUDDLE" IS CREATED WITH A MISAIMED HAMMER, JUST PUT A DAB OF SALIVA IN THE DENT, A DOUBLE DAB IF THE DENT IS DEEP. (IF YOU ARE CHEWING TOBACCO USE A FRIEND).

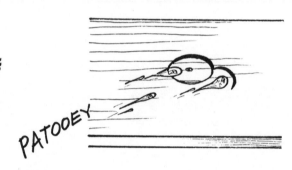

PATOOEY

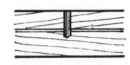 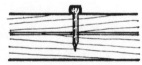 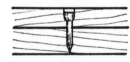

DRILL ... NAIL ... SET

IF THE TRIM IS HARDWOOD, PRE-DRILLING THE NAIL HOLES IS A MUST. A GOOD DRILL BIT IS A NAIL OF THE SIZE BEING USED TO NAIL THE TRIM WITH. CUT THE HEAD OFF AND SHARPEN THE POINT.

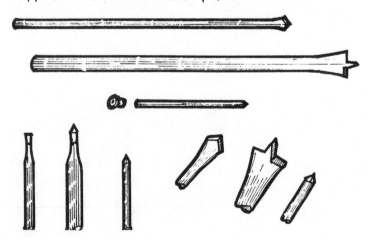

THERE SHOULD NEVER BE A SHORTAGE OF DRILL BITS OF ANY SIZE AS LONG AS THERE IS A PIECE OF STIFF WIRE AROUND. BIG NAILS, LITTLE NAILS, WELDING ROD, ANY-THING WILL DO. BEAT ON THE END TO FLATTEN AND FLAIR THEN FILE CUTTING TIPS.

IF YOU WANT A NAIL TO SLIDE INTO WOOD A LITTLE EASIER, RUB IT IN YOUR HAIR TO PICK UP THE OIL.

BOY THAT FEELS GOOD!!

HUH?

FACE OIL WORKS TOO.

IT ACTS LIKE WAX OR SOAP ON A WOOD SCREW.

PATOOEY

IVORY

SALIVA ON A WOOD SCREW WILL HELP IF SOAP IS NOT AVAILABLE.

WHEN NAILING THIN BRADS, SQUEEZE THE BRAD WITH THE FINGERS AND IT WON'T BEND SO EASILY.

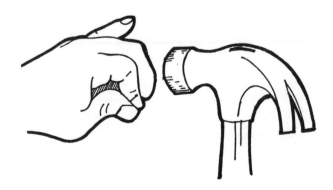

FITTING A SHELF

YOU CAN'T JUST CUT A BOARD SQUARE AT EACH END AND EXPECT IT TO FIT IN A CORNER. THE JOINT COMPOUND BUILD UP CHANGES THE ANGLE IN THE CORNER, SO IT SHOULD BE SCRIBED.

THE SCRIBE TOOL IS JUST A SMALL COMPASS THAT LOCKS AT ANY POSITION DESIRED.

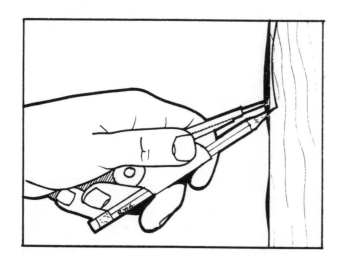

SET THE SCRIBE TO THE DESIRED SPACING AND RUN THE METAL POINT ALONG THE WALL AND WATCH THE PENCIL MAKE A PARALLEL LINE.

MEASURING FOR A SHELF IN A CLOSET IS EASY WITH A FOLDING RULE THAT HAS SLIDING INSETS AT EACH END.

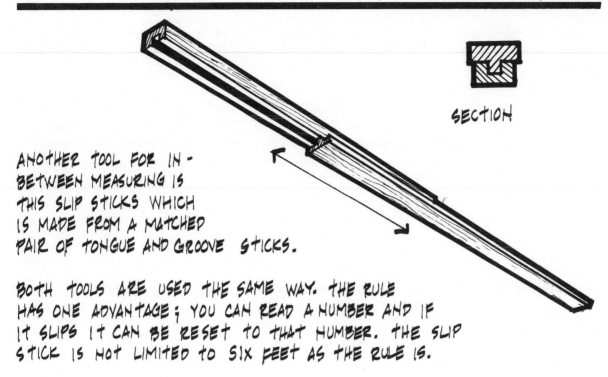

SECTION

ANOTHER TOOL FOR IN-
BETWEEN MEASURING IS
THIS SLIP STICKS WHICH
IS MADE FROM A MATCHED
PAIR OF TONGUE AND GROOVE STICKS.

BOTH TOOLS ARE USED THE SAME WAY. THE RULE
HAS ONE ADVANTAGE; YOU CAN READ A NUMBER AND IF
IT SLIPS IT CAN BE RESET TO THAT NUMBER. THE SLIP
STICK IS NOT LIMITED TO SIX FEET AS THE RULE IS.

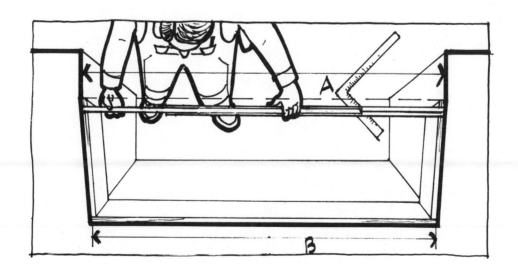

USE THE FOLDING RULE OR SLIP STICK TO FIND THE LONGEST DIMENSION.

FITTING A SHELF

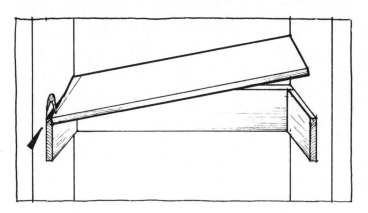

CUT THE BOARD A LITTLE LONGER THAN THE LONGEST DIMENSION, PUT IT IN PLACE TIPPED UP AT ONE END, AND SCRIBE. THE SCRIBE SHOULD BE SET TO HIT THE CORNER OF THE EDGE WHERE THERE IS SPACE BETWEEN THE BOARD AND THE WALL.

CUT THIS LINE STARTING SQUARE AT THE FRONT AND BACK-CUTTING AS YOU PROGRESS. TEST THIS CUT AGAINST THE WALL AND CORRECT WITH A BLOCK PLANE IF NECESSARY.

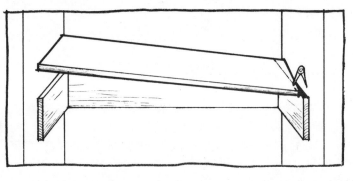

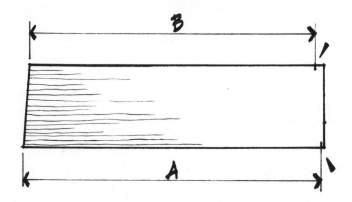

MEASURE THE OPENING, FRONT "A" AND BACK "B", AND TRANSFER THEM TO THE SHELF BOARD.

SLIP THE SHELF IN PLACE, TIPPED UP AT THE OTHER END AND SCRIBE. THE SCRIBE SHOULD BE SET TO HIT ONE OF THE MARKS ON THE BOARD AND WHEN SCRIBED, SHOULD HIT THE OTHER MARK. CUT THIS END THE SAME WAY THE OTHER END WAS CUT AND IT SHOULD BE A PERFECT FIT. A LITTLE BLOCK PLANING MIGHT BE IN ORDER. WITH THE SHELF IN PLACE, SCRIBE AND FIT TO THE BACK WALL.

ALTHOUGH THE FRAME WORK OF STAIRS IS NOT
INSIDE FINISH WORK, IT TIES IN SO CLOSELY
THAT I HAVE TO INCLUDE IT HERE.

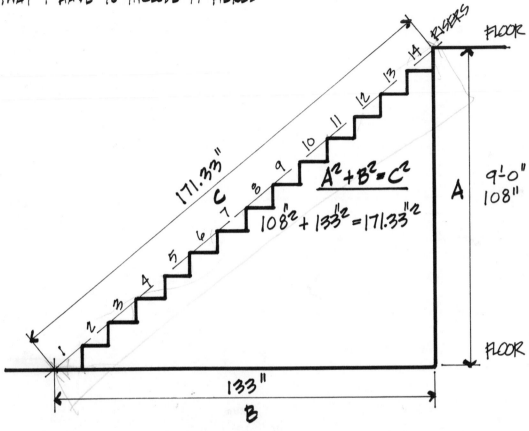

$$A^2 + B^2 = C^2$$
$$108^{\prime\prime2} + 133^{\prime\prime2} = 171.33^{\prime\prime2}$$

THE USE OF THE CALCULATOR MAKES THE MATHEMATICS OF THIS JOB EASY.
TO BEGIN WITH, CHOOSE 14 RISERS (ABOUT AN AVERAGE NUMBER), AND DIVIDE IT
INTO THE FLOOR TO FLOOR HEIGHT (108" IN THIS CASE). 108" ÷ 14 = 7.71" RISER.

$$+ \quad \frac{\text{RISE}}{\text{TREAD}}$$
$$17\frac{1}{2}'' = \text{IDEAL}$$

(FOR THIS EXAMPLE)

THE IDEAL RISE PLUS TREAD TOTAL IS 17½"

	17.50"	IDEAL TOTAL
LESS......	7.71"	THE RISER WE CAME UP WITH
EQUALS....	9.79"	OUR IDEAL TREAD
– USE......	9.50"	A STOCK TREAD SIZE
TIMES......	14	TREAD SPACES
EQUALS...	133"	

FROM SCHOOL, WE REMEMBER $A^2 + B^2 = C^2$, SO $108^2 + 133^2 = 171.3^2$

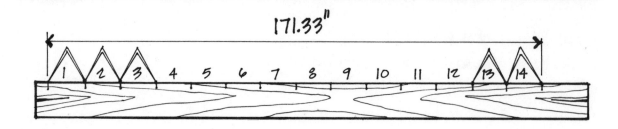

171.33"

MARK THIS DIMENSION ON A 15 FOOT 2×12. IT WILL FIT ON A 14 FOOTER, BUT THE ENDS ARE USUALLY SPLIT AND MAY HAVE OTHER BAD SPOTS WE WOULD LIKE TO AVOID. THIS 171.33" WILL HAVE TO BE DIVIDED INTO THE 14 EQUAL TREAD SPACES WITH A PAIR OF LARGE DIVIDERS.

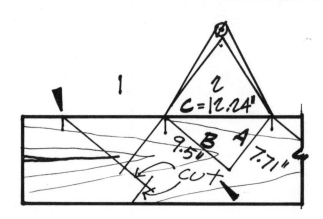

WRITE "CUT" ON THE CARRIAGE SO YOU KNOW WHAT LINE TO CUT.

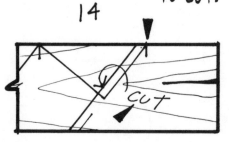

THE EASIEST WAY TO FIND THIS SPACING DIMENSION IS WITH THE SAME $A^2 + B^2 = C^2$ FORMULA.

$$A^2 + B^2 = C^2$$
$$7.71"^2 + 9.5"^2 = 12.24"^2$$

SET THE DIVIDERS AND WALK THEM UP THE 2×12. WITH LUCK YOU WILL HIT THE 171.33" MARK. KEEP ADJUSTING THE DIVIDERS UNTIL YOU GO FROM MARK TO MARK 14 TIMES.

MARK THE 14 SPACES CLEARLY. THEN SET UP A FRAMING SQUARE WITH THE RISE AND TREAD DIMENSIONS BY CLAMPING A 2×4 ACROSS IT.

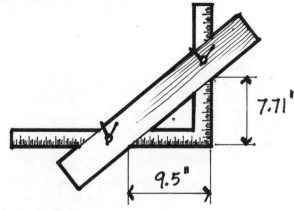

7.71'

9.5"

START AT EITHER END AND
OUTLINE THE RISE AND TREAD
AT EACH MARK.

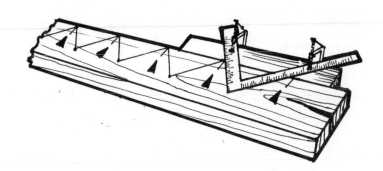

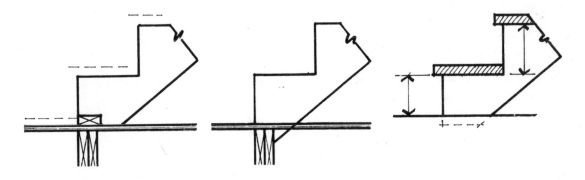

MARK THE BOTTOM OF THIS CARRIAGE FOR WHATEVER FRAMING CONDITION EXISTS.

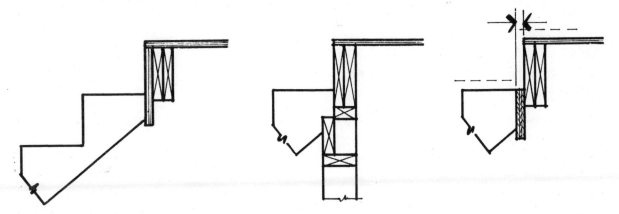

MARK THE TOP OF THE CARRIAGE FOR THE FRAMING CONDITION. DON'T FORGET TO
TAKE THE FINISHED FLOOR INTO ACCOUNT ON BOTH TOP AND BOTTOM.

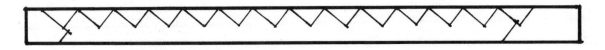

THE CARRIAGE LAYOUT WILL LOOK LIKE THIS.

DOUBLE CHECK THE LAYOUT BEFORE CUTTING. WHEN SATISFIED, FREEHAND CUT WITH A SKIL-SAW OR HAND SAW. RUN THE CUTS A LITTLE PAST WHERE THEY INTERSECT SO THAT THE TRIANGLES WILL FALL OUT. WE CAN USE THESE LATER.

NOW THE CARRIAGE LOOKS LIKE THIS.

TRY IT IN PLACE AND DON'T FORGET THE 3/4" HANGER BOARD IF ONE IS TO BE USED.

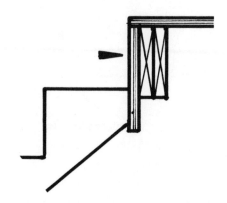

THREE CARRIAGES ARE REQUIRED, SO USE THE COMPLETED CARRIAGE AS A PATTERN FOR THE OTHER 2x12'S. A 2x6 WITH THE TRIANGULAR CUT OUT BLOCKS NAILED ON CAN BE USED FOR THE MIDDLE CARRIAGE.

THE MIDDLE CARRIAGE WITH BLOCKS WILL LOOK LIKE THIS.

IF A HANGER BOARD IS USED, NAIL THE CARRIAGES TO IT AND THEN NAIL THE WHOLE BUSINESS IN PLACE.

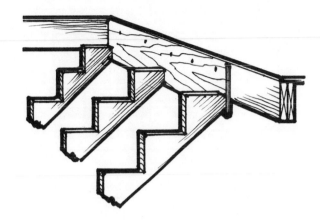

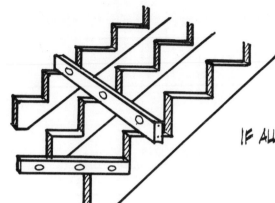

IF ALL IS WELL THE TREADS WILL BE LEVEL.

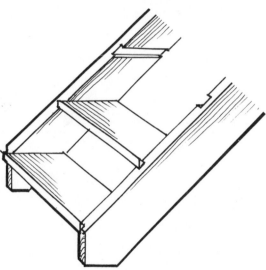

FOR LET-IN TREADS WITH NO RISERS, THE BASIC LAYOUT IS THE SAME.

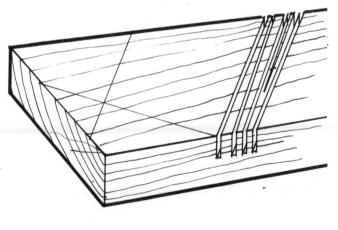

THE GROOVES FOR THE TREADS (USUALLY 2×10.) ARE MADE WITH A SERIES OF PARALLEL CUTS WITH A SKIL-SAW SET TO THE PROPER DEPTH. BREAK AWAY THE PIECES IN THE GROOVE AND CLEAN UP WITH A CHISEL.

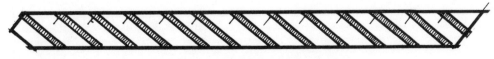

THE FINISHED PRODUCT.

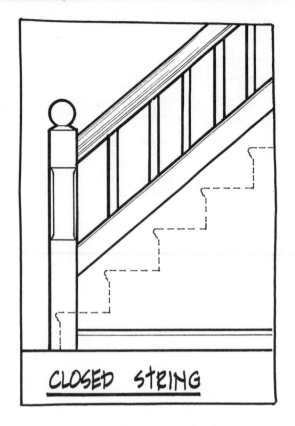

CLOSED STRING

OPEN STRING

FINISHING A STAIRWAY IS A MEASURE-TWICE-CUT-ONCE PRECISION PIECE OF CARPENTRY. THE FIRST HOUSE I BUILT HAD MY FIRST STAIRWAY, SO IT IS NOT IMPOSSIBLE FOR EVEN A BEGINNER. THERE ARE TWO TYPES OF STAIRWAYS: THE CLOSED STRING AND THE OPEN STRING.

THIS IS AN OPEN STRING OR STRINGER, MITERED TO RECEIVE THE RISER.

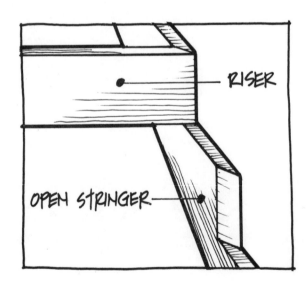

RISER

OPEN STRINGER

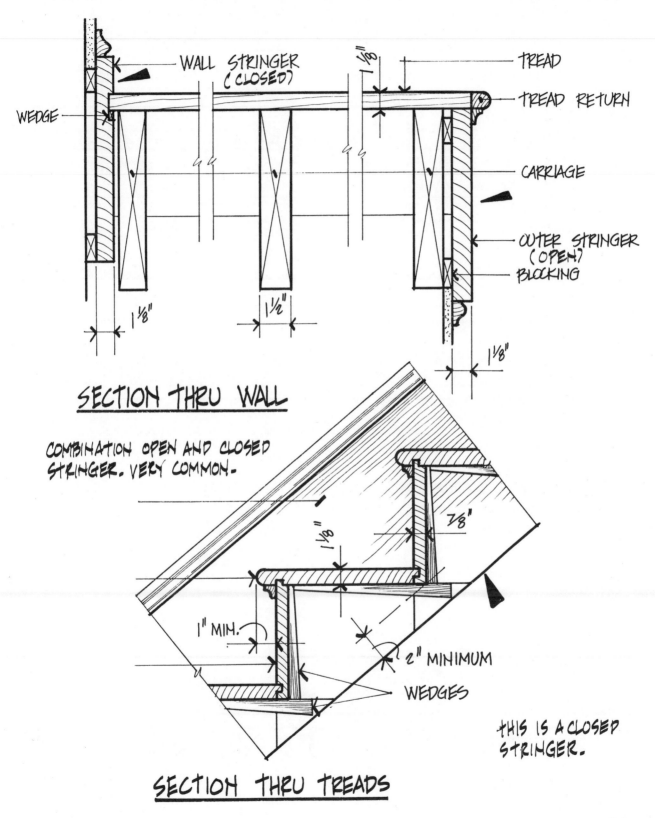

WEDGE

WALL STRINGER (CLOSED)

TREAD

TREAD RETURN

CARRIAGE

OUTER STRINGER (OPEN)

BLOCKING

1/8"

1 1/8"

1 1/2"

1 1/8"

SECTION THRU WALL

COMBINATION OPEN AND CLOSED STRINGER. VERY COMMON.

1 1/8"

7/8"

1" MIN.

2" MINIMUM

WEDGES

THIS IS A CLOSED STRINGER.

SECTION THRU TREADS

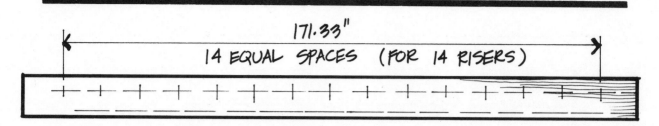

171.33"
14 EQUAL SPACES (FOR 14 RISERS)

THE CLOSED STRINGER IS USUALLY 5/4" STOCK AND IS LAID OUT THE SAME WAY THE CARRIAGE IS LAID OUT, BY DETERMINING THE NUMBER OF RISERS AND THE LENGTH OF TREAD AND RISER.

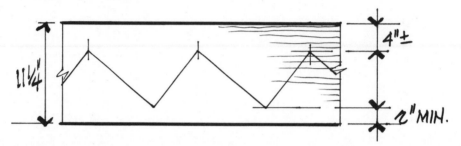

11 1/4"
4"±
2" MIN.

WHEN THE LAYOUT IS MADE, LEAVE AT LEAST 2" CLEARANCE FROM THE BOTTOM OF THE STRINGER TO THE BOTTOM JUNCTION WHERE TREAD AND RISER MEET. THERE SHOULD ALSO BE ABOUT 4" FROM THE TREAD NOSING TO THE TOP OF THE STRINGER. THE TOTAL WIDTH OF THE STRINGER WILL BE 11 1/4"±

SET UP THE FRAMING SQUARE AGAIN, BUT THIS TIME ALLOW FOR THE 2" OFFSET

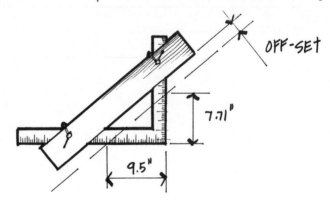

OFF-SET

7.71"

9.5"

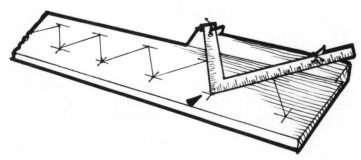

HIT THE MARKS WITH THE FRAMING SQUARE.

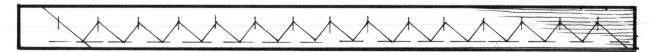

THE RISER AND TREAD LAYOUT WILL LOOK LIKE THIS.

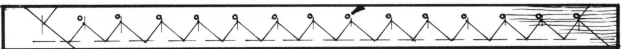

LOCATE AND DRILL 1⅛" NOSING HOLES FOR 1⅛" TREADS SO THAT THERE WILL BE A MINIMUM 1" NOSING OVERHANG PAST THE FACE OF THE RISER.

THIS PLYWOOD GUIDE WILL WORK WITH A KNIFE OR ROUTER. ONE EDGE OF THE GUIDE IS FOR THE TOP OF THE TREAD AND THE OTHER IS FOR THE BOTTOM OF THE WEDGE. CLAMP IN PLACE AND DO YOUR STUFF.

1" MIN

PARALLEL TO WEDGE

EQUAL

½" PLYWOOD

EQ.

EQ

2X4

TREAD

EQUAL

EQ.

2X4

NOSING HOLE

2X4
ALIGNMENT MARK
½" PLYWOOD

THIS GUIDE IS FOR THE RISERS. CLAMP IN PLACE FOR EITHER ROUTER OR KNIFE.

EQUAL - PARALLEL TO WEDGE

EQ. PARALLEL TO RISER   RISER

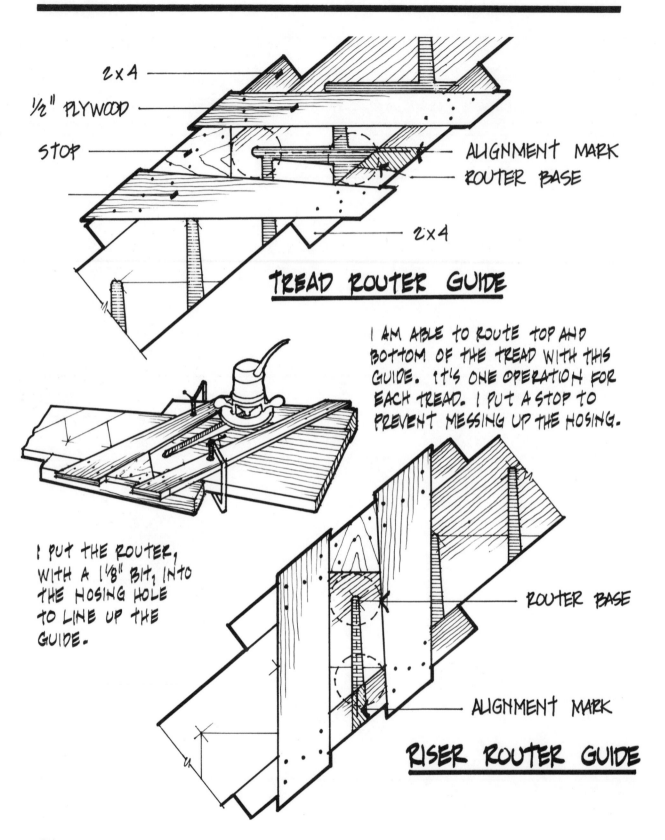

2×4

½" PLYWOOD

STOP

ALIGNMENT MARK
ROUTER BASE

2×4

TREAD ROUTER GUIDE

I AM ABLE TO ROUTE TOP AND
BOTTOM OF THE TREAD WITH THIS
GUIDE. IT'S ONE OPERATION FOR
EACH TREAD. I PUT A STOP TO
PREVENT MESSING UP THE HOSING.

I PUT THE ROUTER,
WITH A 1⅛" BIT, INTO
THE HOSING HOLE
TO LINE UP THE
GUIDE.

ROUTER BASE

ALIGNMENT MARK

RISER ROUTER GUIDE

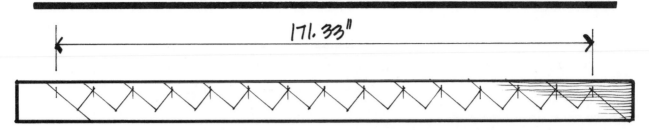

171.33"

THE OPEN STRINGER IS LAID OUT THE SAME WAY TO START.

THE FACE OF THE RISER IS LOCATED, IN THIS CASE, 7/8" FROM THE RISER LAYOUT LINE. THIS IS THE OUTSIDE CORNER OF THE MITER CUT.

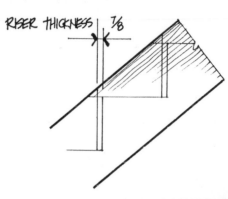

RISER THICKNESS 7/8

THE STRINGER IS ALL LAID OUT AND READY TO BE CUT WITH A SKIL-SAW AND GUIDE OR A HAND SAW. I PREFER THE HAND SAW.

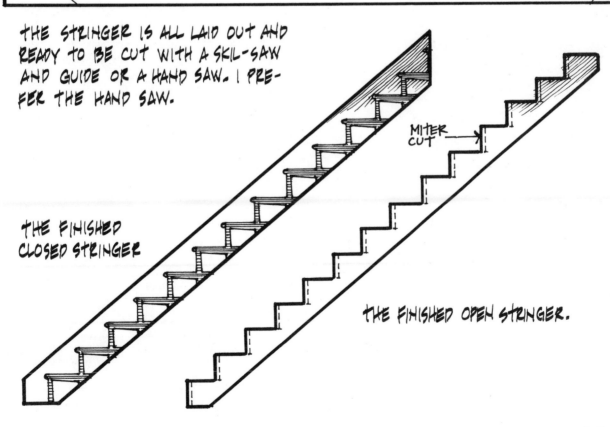

MITER CUT

THE FINISHED CLOSED STRINGER

THE FINISHED OPEN STRINGER.

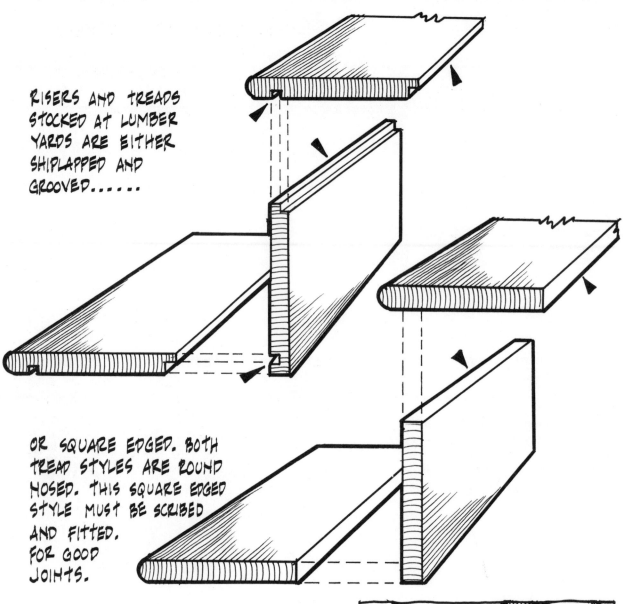

RISERS AND TREADS
STOCKED AT LUMBER
YARDS ARE EITHER
SHIPLAPPED AND
GROOVED......

OR SQUARE EDGED. BOTH
TREAD STYLES ARE ROUND
NOSED. THIS SQUARE EDGED
STYLE MUST BE SCRIBED
AND FITTED.
FOR GOOD
JOINTS.

THE JOINT WHERE RISER AND TREAD
INTERLOCK (FRONT AND BACK), WEDGES
AND BLOCKS, ARE GLUED AS YOU PRO-
GRESS. THE WEDGES ARE DRIVEN IN
TO FORCE THE FACE OF THE TREAD AND
RISER AGAINST THE FACE OF THE ROUTED
STRINGER. DON'T SPARE THE GLUE.

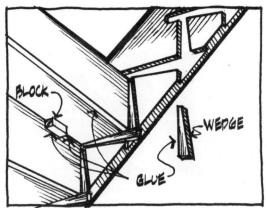

BLOCK

WEDGE

GLUE

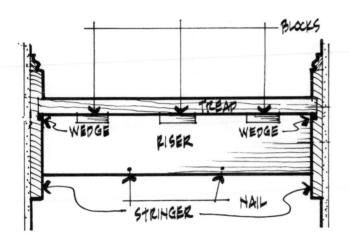

THE ROUTED CLOSED STRINGER STAIRWAY DOES NOT HAVE CARRIAGES, BECAUSE IT IS ASSEMBLED FROM THE UNDERSIDE. START AT THE TOP INSTALLING THE TREAD FIRST.

THE SCRIBED AND BUTTED CLOSED STRINGER DOES HAVE CARRIAGES AND IT IS ASSEMBLED FROM THE TOPSIDE STARTING FROM THE TOP, PUTTING IN THE TREAD FIRST.

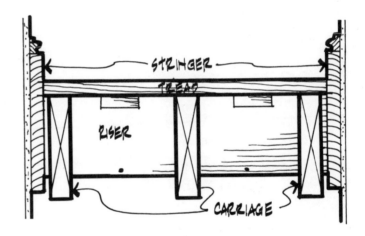

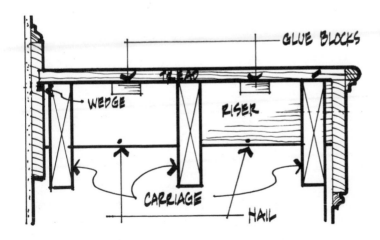

THE COMBINATION OPEN AND CLOSED STRINGER USES CARRIAGES AND IS ASSEMBLED FROM THE TOPSIDE STARTING FROM THE BOTTOM, PUTTING THE RISERS IN FIRST.

HARDWOOD FLOORING

NAILING HARDWOOD FLOORING IS GREATLY EASED WITH THE USE OF THE NAILING MACHINE. THESE MACHINES CAN BE RENTED AT MOST LUMBER YARDS.

THE OLD FLOORING HAMMER WILL ALSO DO THE TRICK. THIS HAMMER IS ABOUT 25 OZ WITH A LITTLE BIGGER HEAD AND A LONGER HANDLE. THE HEAD IS ALSO A HARDER STEEL THAN A STANDARD HAMMER TO HANDLE THE HARD FLOORING NAILS.

NO MATTER WHAT TOOL YOU USE, THE BENT OVER POSITION IS THE ONE ASSUMED FOR THE WHOLE JOB.

IT'S BEST TO HAVE THE FLOORING STOCK ON THE JOB A FEW DAYS IN THE HEATED HOUSE, PREFERABLY UNBUNDLED, SO THAT THE STRIPS CAN ADJUST TO THE CLIMATE.

THE STRIPS ARE BEST LAID RUNNING THE LONG WAY IN THE ROOM.

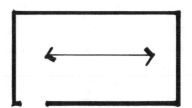

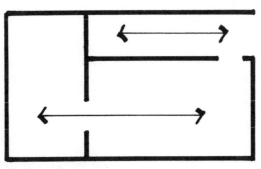

WHEN A ROOM OPENS TO ANOTHER ROOM OR HALL, KEEP THE STRIPS RUNNING IN THE SAME DIRECTION

THE STRIPS CAN GO EITHER WAY WITH DIAGONAL SUB-FLOORING, BUT THE STRONGEST CONDITION IS TO RUN THE STRIPS PERPENDICULAR TO THE JOISTS. WITH DIAGONAL SUB-FLOORING THIS IS ALWAYS POSSIBLE.

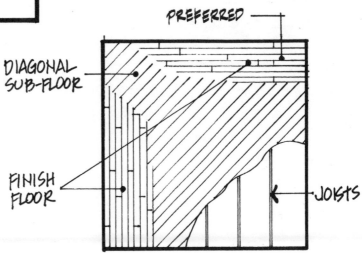

WITH THE SUB-FLOOR RUNNING PERPENDICULAR TO THE JOISTS, THE FINISHED FLOOR SHOULD RUN PERPENDICULAR TO THE SUB-FLOOR FOR THE STRONGEST CONDITION.

HARDWOOD FLOORING

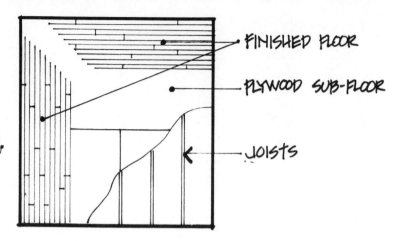

FINISHED FLOOR

PLYWOOD SUB-FLOOR

JOISTS

WITH PLYWOOD SUB-FLOORING, THE STRONGEST CONDITION IS WITH THE STRIPS RUNNING PERPENDICULAR TO THE JOISTS, BUT EITHER WAY IS O.K.

THE FIRST THING TO DO BEFORE STARTING THE FLOORING IS A CLEAN SWEEP.

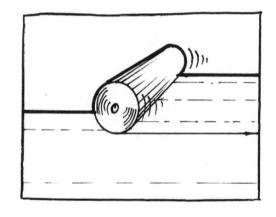

THEN ROLL OUT 15# BUILDING PAPER ON THE FLOOR BEING LAID. LAP THE EDGES A FEW INCHES TO KEEP DUST, DIRT AND DAMP FROM THE FLOORING.

TO ESTABLISH THE LOCATION OF THE FIRST STRIP OF FLOORING, SNAP A CHALK LINE OR STRETCH A STRING 6 OR 8 INCHES FROM THE WALL AND PARALLEL. THE STRING IS BETTER BECAUSE THE PAPER IS LIABLE TO MOVE, CHANGING THE CHALK LINE LOCATION. START THE FIRST STRIP ½" FROM THE WALL AND KEEP IT PARALLEL TO THE STRING OR CHALK LINE.

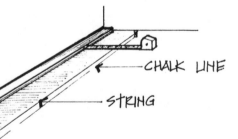

CHALK LINE

STRING

92

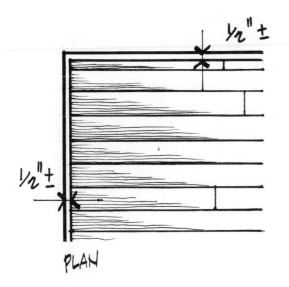

½" ±

½" ±

PLAN

THE FLOORING IS KEPT ABOUT ½" AWAY FROM THE WALL ALL AROUND TO ALLOW FOR SWELLING.

THE BASE OR SHOE BASE COVERS THIS SPACE.

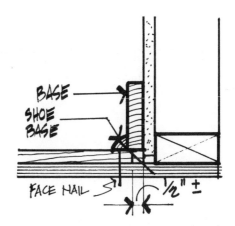

BASE

SHOE BASE

FACE NAIL

½" ±

6" to 8"

FACE NAIL THIS FIRST PIECE USING A CUT-OFF FINISH NAIL TO PRE-DRILL. LAY OUT A FEW ROWS AHEAD SO YOU JUST HAVE TO MOVE THE STRIPS INTO PLACE AND NAIL THEM HOME. STAGGER THE JOINTS 6 TO 8 INCHES.

HARDWOOD FLOORING

THE SKILL REQUIRED WITH THE HAMMER IS NOT NECESSARY WITH THE MACHINE.

THE BASIC WORKING POSITION FOR HAMMER AND MACHINE.

THE NAILING IS DONE BETWEEN THE FEET.

DRIVE THE NAIL ALMOST HOME, JUST SHORT OF HITTING THE FLOORING WITH THE HAMMER.

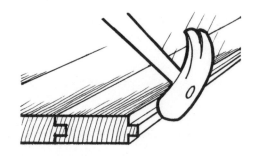

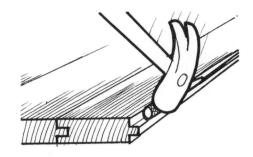

LAY A ROUND NAIL SET ON THE NAIL HEAD AND DRIVE HOME WITH THE HAMMER BEING CAREFUL NOT TO HIT THE FLOORING.

THE MACHINE, LOADED WITH NAILS, IS PLACED OVER THE EDGE OF THE FLOORING AND POUNDED WITH A HEAVY MALLET. IT'S THE WAY TO GO.

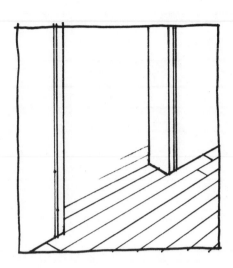

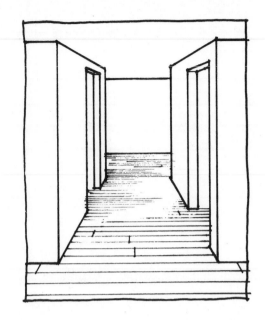

MIX SHORT PIECES WITH LONG ONES, BUT IN HALLS AND AT DOORWAYS WHERE THERE IS A LOT OF TRAFFIC, USE LONG PIECES. USE EXTRA SHORT AND BAD LOOKING PIECES IN CLOSETS.

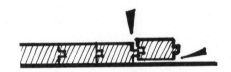

STAND ON THE STRIP GOING IN AND THE PREVIOUS ONE TO ALIGN AND TO HOLD IT IN PLACE.

USE A BLOCK OF FLOORING MATERIAL TO TAKE THE HAMMER BLOWS IF A PIECE OF FLOORING IS TOUGH TO GET IN TIGHTLY. THE NAILING MACHINE ACTS THE SAME WAY WHEN IT'S POUNDED WITH THE HEAVY MALLET.

HARDWOOD FLOORING

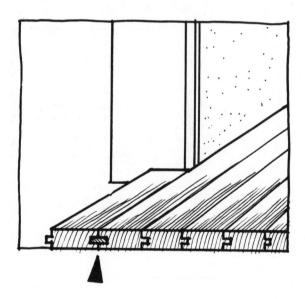

WHEN WORKING FROM A WALL ALONG SIDE A CLOSET, LAY THE FLOOR UP TO THE CLOSET FACE AND REVERSE THE FLOORING GOING INTO THE CLOSET. WHEN YOU REVERSE, THE STRIPS WILL BE GROOVE TO GROOVE. A SPLINE CUT TO FIT WILL TAKE CARE OF THAT. THE REST OF THE FLOORING GOES IN ONE DIRECTION WHILE THE CLOSET FLOORING GOES IN THE OTHER.

THE LAST PIECE OF FLOORING IS RIPPED, DRILLED, AND WEDGED INTO POSITION WITH A PRY BAR BEFORE NAILING. BE SURE TO PROTECT THE WALL OR BASE BOARD WITH A BLOCK OF WOOD.

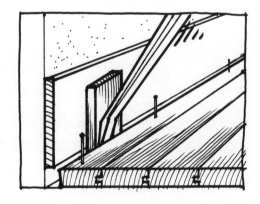

AN OLD TIME CABINET MAKER TOLD ME TO
ALWAYS USE PLYWOOD LEDGER STRIPS FOR
SUPPORTING COUNTER TOPS BECAUSE IT
WON'T SHRINK.

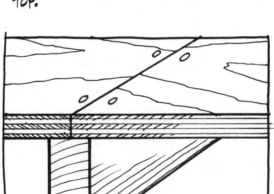

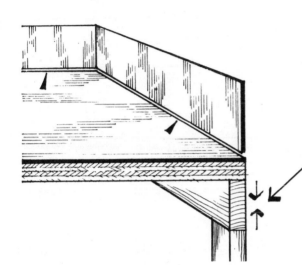

SOLID STOCK WILL SHRINK, CAUSING A GAP
BETWEEN THE BACKSPLASH AND THE COUNT-
ER TOP.

ALL PLYWOOD SEAMS SHOULD BE WELL
BLOCKED UNDERNEATH.

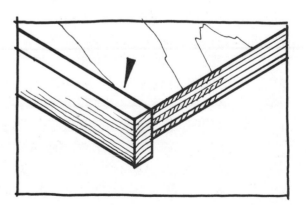

I LIKE THIS KIND OF NOSING; THE TOP IS
MORE SOLID.

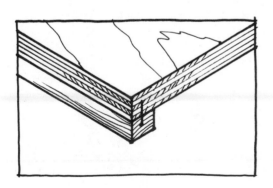

NOSING APPLIED TO THE FACE OF THE
COUNTER TOP CAN LOOSEN, DISTURBING
THE FORMICA CORNER.

COUNTER TOPS

FILL ALL SEAMS AND DEPRESSIONS. SET AND FILL ALL NAIL HOLES AND FINALLY, SAND THE WHOLE BUSINESS.

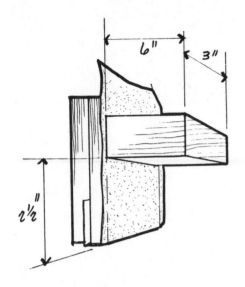

MAKE UP A SIMPLE JIG FROM SCRAP WOOD TO SAND THE EDGES.

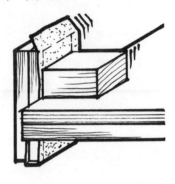

I HAVE PUT IN MANY FORMICA COUNTER TOPS USING A SAW, CHISEL, BLOCK PLANE AND A FILE, BUT I NOW HAVE A FEW SIMPLE TOOLS TO MAKE THE JOB EASIER. THE QUALITY DOESN'T CHANGE WITH THESE TOOLS, BUT THE SPEED DOES.

THE SIMPLEST TOOL FOR CUTTING FORMICA (PLASTIC LAMINATE) IS A KNIFE WITH A CARBIDE TIP.

PUT A STRAIGHT EDGE ON THE FORMICA SHEET AND SCORE WITH THE KNIFE. IT MIGHT TAKE A FEW STROKES. THEN SNAP ALONG THE SCORED LINE. PUT THE STRAIGHT EDGE ON THE GOOD SIDE OF THE LINE. IF THE KNIFE RUNS OFF THE LINE, THE GOOD PIECE WONT BE RUINED. YOU CAN EVEN CUT HOLES FOR SWITCHES AND SINKS WITH THIS TOOL. IT HAS TO BE USED WITH CARE BECAUSE IT DOESN'T CONTROL TOO EASILY (IT IS DIFFICULT TO START AND STOP THE CUTS). A FEW PRACTICE CUTS WILL HELP YOUR CONFIDENCE.

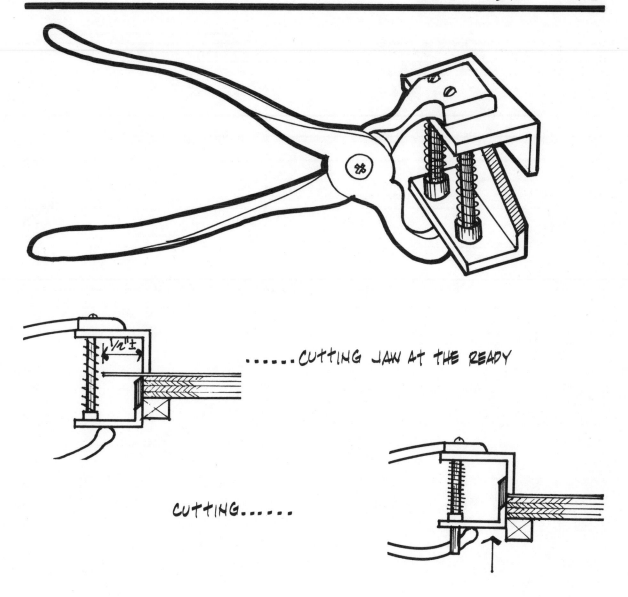

...... CUTTING JAW AT THE READY

CUTTING ......

ANOTHER GREAT TOOL IS THIS PAIR OF SHEARS. AFTER THE FORMICA IS INSTALL-
ED THESE SHEARS WILL CUT RIGHT TO THE EDGE WITH A CLEAN CUT. THE SHEET
SHOULD HAVE NO MORE THAN ½" OVERHANG. THIS TOOL ALSO CUTS RIGHT UP
TO THE WALL, UNLIKE A ROUTER THAT LEAVES A FEW INCHES TO CHISEL.
THE KNIFE CUTS THE SHEETS TO THE APPROXIMATE SIZE AND THEN THE SHEARS
TRIM THEM TO THE EXACT SIZE AFTER THEY ARE CEMENTED IN PLACE. A FILE
IS USED TO EASE THE SHARP CORNER.

COUNTER TOPS

THE EDGING IS PUT ON BEFORE THE TOP. 1½" EDGING PIECES ARE AVAILABLE, OR THEY CAN BE CUT WITH A KNIFE OR ON THE TABLE SAW USING A PLYWOOD CUTTING BLADE.

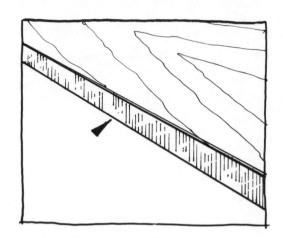

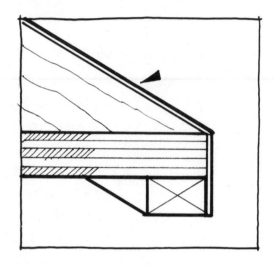

IT CAN EASILY BE PUT ON FLUSH WITH THE TOP. USE A GOOD EDGE UP FOR A TIGHT CORNER

IT CAN BE PUT ON WITH A LITTLE OVERHANG AND TRIMMED LATER WITH ROUTER OR SHEARS.

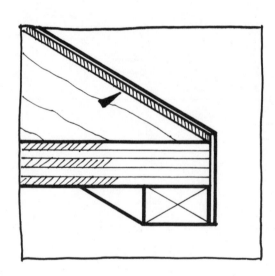

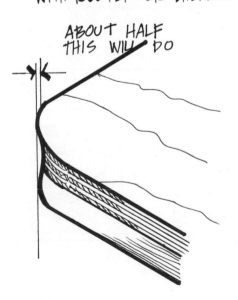

ABOUT HALF THIS WILL DO

ROUNDED CORNERS ARE EASY IF A LITTLE CARE IS TAKEN WHEN CUTTING THE PLYWOOD TOP. I LIKE TO CONCAVE THE FACE A LITTLE TO BE SURE THE TOP AND BOTTOM OF THE FORMICA EDGING HAVE GOOD CONTACT. WHEN THE FORMICA EDGING IS BENT, THE FACE WILL BE SLIGHTLY CONCAVE.

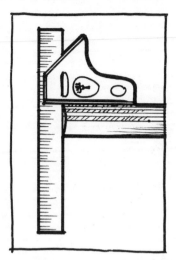

THE EDGE MUST ALSO BE SQUARE WITH THE TOP.

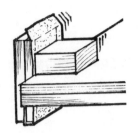

THIS JIG WILL HELP

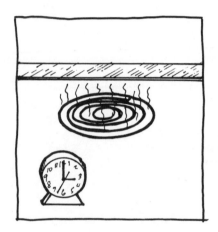

FORMICA WILL FOLLOW LARGE CURVES WITH NO TROUBLE, BUT SMALL CURVES NEED HELP. HEAT IS THE HELP NEEDED AND A HOT PLATE OR STOVE WILL DO THE JOB. FIRST, TEST A SAMPLE PIECE OF EDGING BY RESTING IT A FEW INCHES FROM THE HEAT AND TIMING HOW LONG IT TAKES TO "POP". NOW TAKE A GOOD PIECE, HEAT IT JUST SHORT OF THAT TIME...

AND BEND IT AROUND THE CURVE. THERE SHOULD BE NO CEMENT ON THE EDGING OR THE COUNTER FACE. YOU ARE JUST PRE-FORMING THE EDGING. REMEMBER, SOME CONTACT CEMENT IS VERY FLAMABLE, SO BE CAREFUL. THE PRE-FORMED EDGE CAN THEN CEMENTED IN PLACE. ANOTHER WAY IS TO THIN THE STRIP ON THE BACK SIDE, WITH A BLOCK PLANE, WHERE THE CURVE IS TO BE. THIS SYSTEM HAS LIMITATIONS.

COUNTER TOPS

A FILE LAID FLAT ON THE COUNTER TOP AND RUN ALONG THE EDGE WILL TAKE CARE OF ANY HIGH SPOTS THE EDGING MIGHT HAVE.

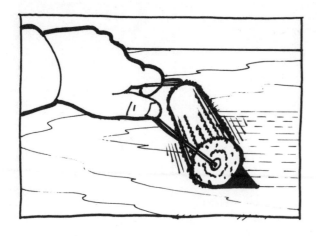

THE EASIEST AND QUICKEST WAY TO APPLY CONTACT CEMENT IS WITH A PAINT ROLLER.

THE FUMES WILL MAKE YOU LIGHT HEADED IN A HURRY, SO VENTILATE.

ON A LONG COUNTER, AND WHEN MORE THAN ONE PIECE OF FORMICA IS TO BE PUT DOWN, 1/4" STRIPS OF WOOD ABOUT 12" APART WILL KEEP THE CEMENT COATED SHEETS OFF THE CEMENT COATED PLYWOOD SO THAT THEY CAN BE POSITIONED.

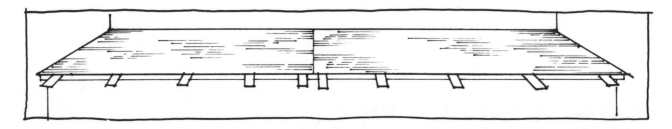

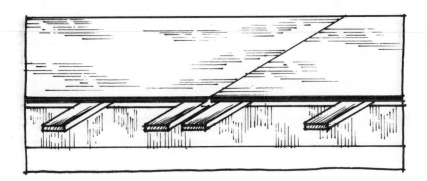

PUT A STRIP ON EACH
SIDE OF THE SEAM.

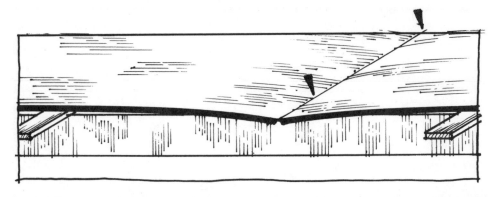

PULL THE STRIPS AT THE SEAM FIRST, MAKING SURE THE SEAM IS PERFECT. THEN PULL THE STRIPS FARTHEST AWAY AND WORK TOWARD THE SEAM. THIS WILL DRIVE THE TWO SHEETS TOGETHER, MAKING A TIGHT FIT AT THE SEAM.

THE SAFEST WAY TO CEMENT AN ODD SHAPED BACKSPLASH THAT TUCKS UP UNDER CABINETS IS WITH WALLBOARD MASTIC. IT ALLOWS YOU TO MOVE THE SHEET AROUND ON THE WALL. CONTACT CEMENT IS GOOD IF THERE ARE ENOUGH HANDS TO HOLD THE SHEET OFF THE WALL WHILE IT IS BEING POSITIONED.

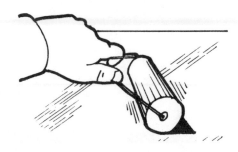

ONCE THE SHEETS ARE CEMENTED IN PLACE
THEY SHOULD BE ROLLED WITH A 6" HARD RUBBER
ROLLER TO GET THE AIR OUT AND MAKE A
GOOD BOND.

A BLOCK OF WOOD AND A HAMMER WORKS TOO.
BE SURE TO COVER THE WHOLE SURFACE.

COUNTER TOPS

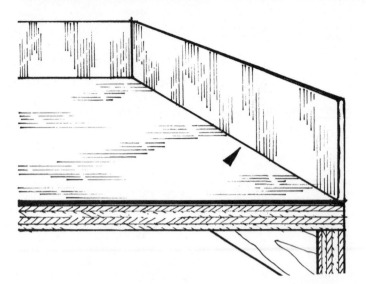

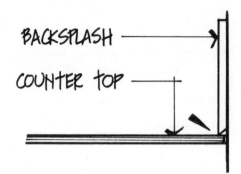

BACKSPLASH

COUNTER TOP

COUNTER TOPS GO ON BEFORE THE BACKSPLASH AND, LIKE CLOSET SHELVES, THE EDGES THAT RUN ALONG THE WALLS HAVE TO BE SCRIBED. IT DOESN'T HAVE TO BE A SUPER FIT BECAUSE THE BACKSPLASH WILL COVER ANY IMPERFECTIONS. THE BACKSPLASH DOES HAVE TO BE A GOOD SCRIBE FIT.

A GOOD TOOL FOR THIS IS A SHARP BLOCK PLANE. LIKE ANY GOOD BUTT JOINT, PLANE IT SO THAT THE FACE EDGE IS STRONG. AS LONG AS THE PLANE IS SHARP, IT WORKS SIMILAR TO PLANING WOOD, BUT WHEN IT DULLS (AND IT DULLS QUICKLY) THE WORK GETS DIFFICULT. THE PROCEDURE IS; SCRIBE, PLANE, FIT, (ALMOST), SCRIBE, PLANE, FIT (HOPEFULLY). TO PLANE FORMICA, HOLD THE SHEET, WITH JUST A LITTLE OVERHANG, AT THE EDGE OF THE COUNTER. HOLD THE SHEET DOWN WITH ONE HAND WHILE MAKING CLEAN, SURE STROKES WITH THE PLANE. TRY LONGISH STROKES RATHER THAN SHORT CHOPPY ONES. KEEP THE PLANE SHARP.

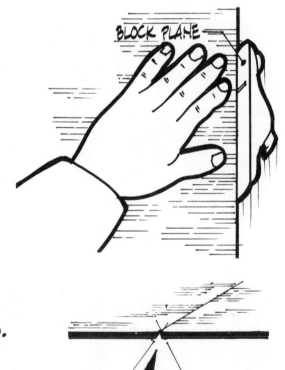

BLOCK PLANE

WHEN BUTTING PIECES USE THE SAME BACK CUTTING TECHNIQUE TO ENSURE A TIGHT FIT ON TOP.

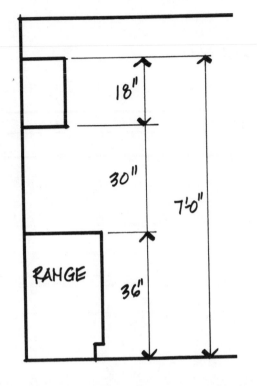

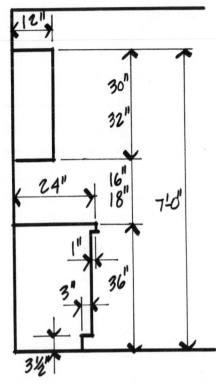

IF KITCHEN CABINETS ARE TO BE HOME MADE (CUSTOM) ANY SIZE WILL DO, BUT THERE ARE SOME BASIC STANDARD DIMENSIONS.

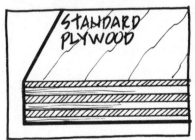

STANDARD PLYWOOD

ANOTHER CONSIDERATION IS THE MATERIAL. AGAIN, ANYTHING WILL DO, BUT THE BEST CHOICE IN PLYWOOD IS LUMBER CORE. IT'S EXPENSIVE, SO USE ONLY WHERE NECESSARY. IT'S MADE UP OF THREE LAYERS, A VENEER ON EACH SIDE OF SOLID STRIPS OF WOOD.

LUMBER CORE

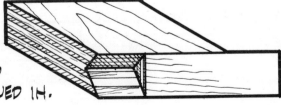

THE EDGES CAN BE COVERED WITH SOLID STRIPS OR A TRIANGULAR PIECE CAN BE GLUED IN.

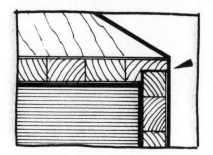

USE THE RABBET JOINT WHENEVER POSSIBLE. IT'S STRONG AND NEAT.

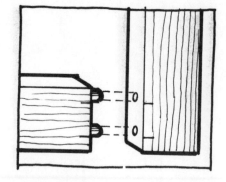

DOWELED JOINTS CAN'T BE BEAT

THESE ARE MINIMUM LENGTHS. USE LONGER SCREWS WHEN WORK PERMITS AND PRE-DRILL. GLUE JOINTS WHERE POSSIBLE.

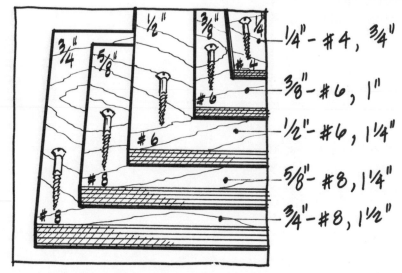

1/4" - #4, 3/4"

3/8 - #6, 1"

1/2" - #6, 1 1/4"

5/8" - #8, 1 1/4"

3/4" - #8, 1 1/2"

PRE-DRILL IF NAILING IS CLOSE TO THE EDGE.

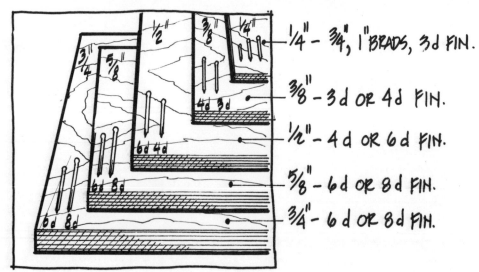

1/4" - 3/4", 1"BRADS, 3d FIN.

3/8 - 3d OR 4d FIN.

1/2" - 4d OR 6d FIN.

5/8" - 6d OR 8d FIN.

3/4" - 6d OR 8d FIN.

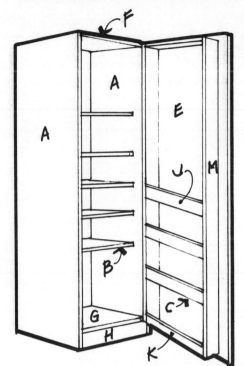

BEFORE STARTING CONSTRUCTION ON CABINETS, THERE SHOULD BE A DRAWING OF SOME SORT THAT ALL THE PARTS CAN BE LETTER CODED ON.

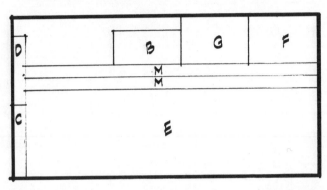

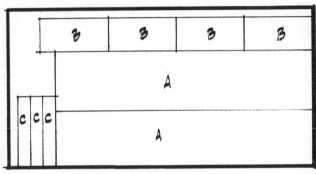

THEN, ON GRAPH PAPER, LAY OUT A 4 × 8 PLYWOOD SHAPE (ABOUT 1½" 1'-0") AND FIT THE PARTS IN. KEEP THE GRAIN DIRECTION AND SAW CUT WIDTH IN MIND. THE NUMBER OF SHEETS REQUIRED FOR THE JOB IS EASY TO FIGURE FROM THESE LAYOUTS. THEY DON'T HAVE TO BE FANCY.

THE DEPTH OF THE BASE IS CONTROLLED BY THE CABINET FRONT FRAME.

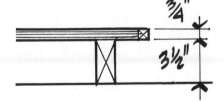

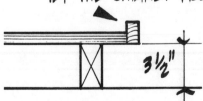

---- THIS CONDITION IS HARD TO CLEAN.

THIS IS FAIRLY COMMON . . . . .

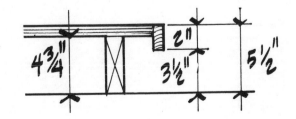

CABINETS

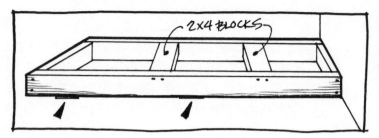

2X4 BLOCKS

THERE ARE MANY WAYS TO BUILD CABINETS AND I LIKE TO BUILD BASE CABINETS ON A LEVELED 1½"x 4¾" BASE. LEVEL UP WITH WOOD SHINGLE TIPS. THE BASE IS DESIGNED SO THAT THE INTERIOR PARTITIONS CAN BE CENTERED ON 2x4 BLOCKS BUILT INTO THE BASE.

PENCIL THE CABINET OUTLINE ON THE WALL. NAIL THE 3/4" PLYWOOD COUNTER SUPPORT 3/4" DOWN FROM THE LEVEL COUNTER TOP LINE. BE SURE TO FIND THE STUDS WITH THE NAILING. NOW IT'S READY TO BUILD ON.

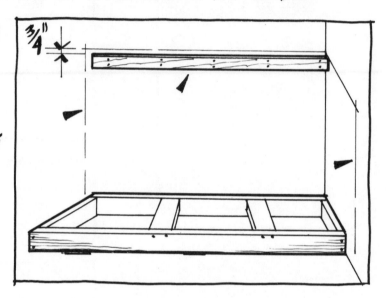

3/4"

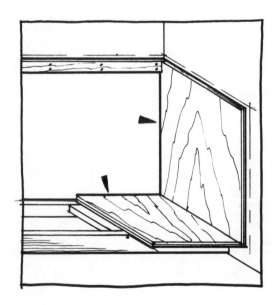

WHETHER THE CABINETS ARE CUT AND FIT AS YOU GO OR PRE-CUT, THIS SYSTEM WORKS NICELY.
I START WITH THE WALL PIECE THAT IS NOTCHED TO FIT SNUGLY UNDER THE WALL RAIL PIECE (COUNTER SUPPORT). KEEP THIS PIECE 3/4" FROM THE WALL MARKS. THE FLOOR PIECE FOLLOWS. IF THE BASE IS LEVEL, ALL SHOULD GO WELL.

PLUMB THE PARTITIONS AS YOU GO.

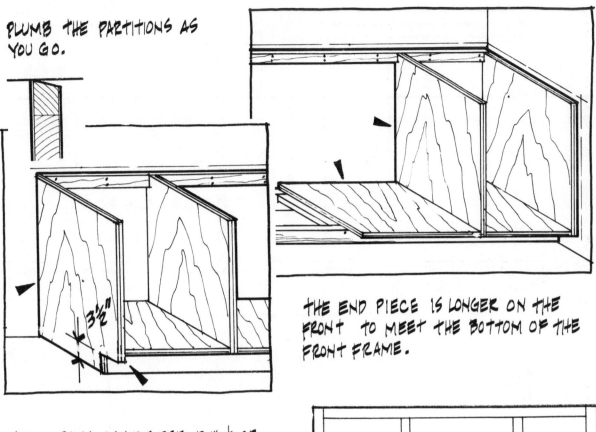

3½"

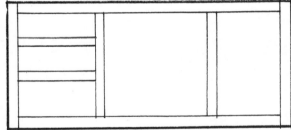

THE END PIECE IS LONGER ON THE FRONT TO MEET THE BOTTOM OF THE FRONT FRAME.

THE FRAME CAN BE PRE-BUILT OR CUT AND FIT. IF PRE-BUILT, DOWEL AND GLUE JOINTS. IF CUT AND FIT, GLUE AND NAIL JOINTS

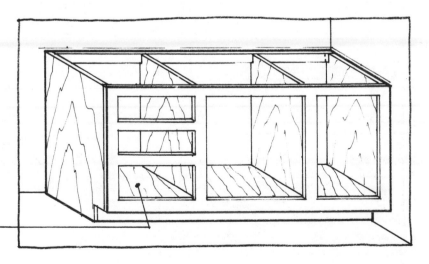

TO ALLOW FOR SCRIB-ING, MAKE THE FRAME A TAD WIDER AND BEVEL THE EDGE THAT HITS THE WALL.

THIS PANEL IS NOT NECESSARY IF THERE IS A DRAWER HERE

CABINETS

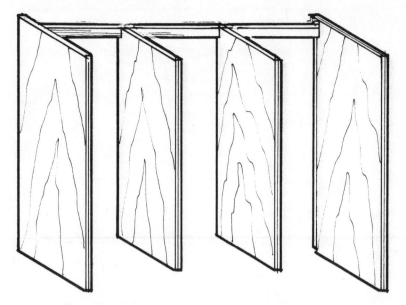

I BUILD THE UPPER CABINETS WITH INTERIOR PARTITIONS LIKE THE BASE CABINETS. NOTE THAT THE MIDDLE PARTITIONS ARE SHORTER, TO SIT ON TOP OF THE BOTTOM PIECE.

THE TOP AND BOTTOM ARE ALSO ONE PIECE SECTIONS.

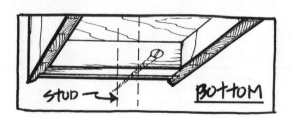

THERE IS A CLEAT AT THE TOP AND BOTTOM FOR FASTENING THE CABINET TO THE WALL.

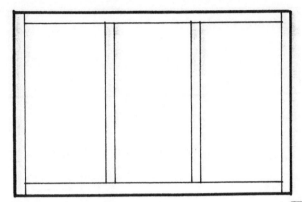

IF THE CABINET STARTS AT A WALL, MAKE THE SIDE RAIL WIDE AND BEVELED TO ALLOW FOR SCRIBING.

HERE AGAIN, THE FRONT FRAME CAN BE PRE-BUILT OR CUT AND FIT AS YOU GO.

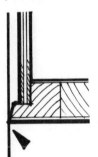

THE SIDE PANELS ARE RABBETED AND BEVELED FOR SCRIBING.

DOORS CAN BE SQUARE EDGED OR LIPPED.

DOOR   DOOR

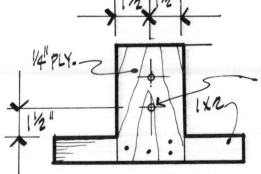

1½" 1½"

¼" PLY.

1½"

THIS HOLE FOR SINGLE KNOB

1×2

A JIG FOR LOCATING DOOR PULL HOLES WILL SPEED THINGS UP.

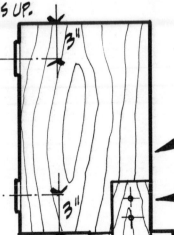

3"

3"

MAKE IT SO THAT WHEN HELD TIGHT TO THE BOTTOM OR TOP AND FLUSH TO THE DOOR EDGE, THE HOLES WILL BE RIGHT. IT IS BOTH A LEFT AND A RIGHT JIG.

CABINETS

TO HANG WALL CABINETS, A COUPLE OF WOOD HELPERS MAKE THE JOB EASY. THEY SHOULD BE A LITTLE SHORT SO THAT THE CABINETS CAN BE SHIMMED TO THE RIGHT HEIGHT.

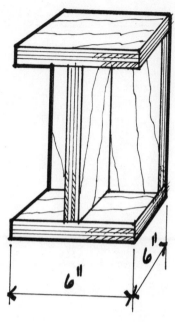

WHEN EVERYTHING IS LINED UP, CHECK THE FIT AT THE WALL. SCRIBE AND PLANE IF NECESSARY.

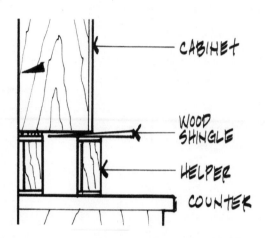

CABINET

WOOD SHINGLE

HELPER

COUNTER

6" 6"

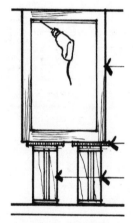

IF ALL IS WELL, FIND THE STUDS, DRILL AND SCREW. IT WILL TAKE A 3" SCREW, SO HAVE SOME SOAP ON HAND.

DRAWERS SHOULD BE SOLID AND EASY TO OPERATE. THE SMOOTHEST OPERATING SLIDES ARE STORE BOUGHT OF WHICH THERE ARE MANY TYPES. A SIMPLE SYSTEM IS TO EXTEND A 1/2" PLYWOOD BOTTOM TO THE SIDES AND RUN IT IN GROOVES IN THE SIDE PANELS. USE PLENTY OF PARAFFIN OR SILICONE TO HELP THE ACTION.

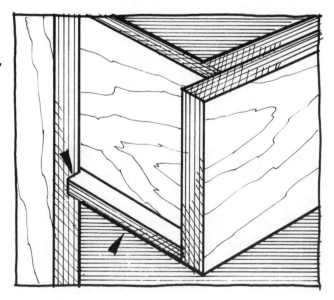

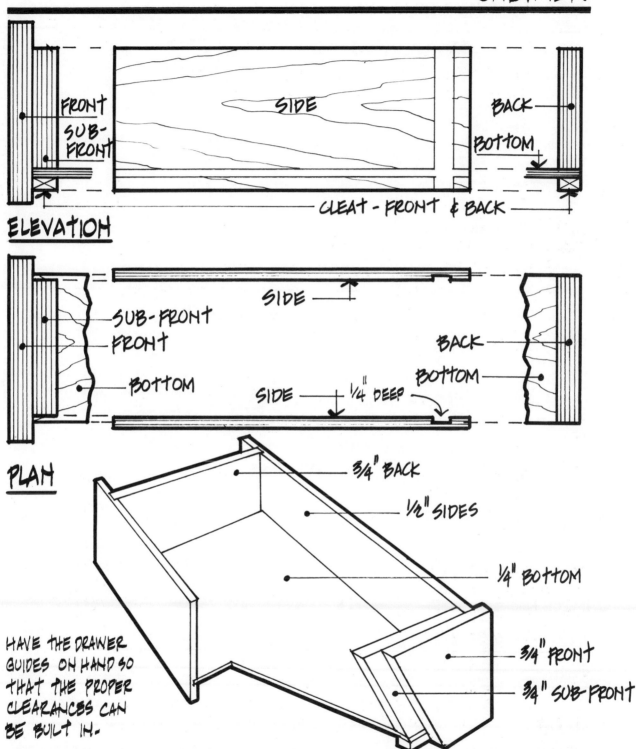

ELEVATION

FRONT
SUB-FRONT
SIDE
BACK
BOTTOM
CLEAT - FRONT & BACK

PLAN

SUB-FRONT
FRONT
BOTTOM
SIDE
SIDE
¼" DEEP
BACK
BOTTOM

¾" BACK
½" SIDES
¼" BOTTOM
¾" FRONT
¾" SUB-FRONT

HAVE THE DRAWER
GUIDES ON HAND SO
THAT THE PROPER
CLEARANCES CAN
BE BUILT IN.

THE CABINETS WILL GO TOGETHER BETTER IF ALL THE PIECES OF THE SAME DIMENSION
ARE CUT WITH THE SAME SAW SETTING. IN THIS CASE, THE SUB-BASE AND BACK
ARE THE SAME HEIGHT, THE BOTTOM AND BACK ARE THE SAME WIDTH.

BOOKCASES

SAP SIDE

HEART SIDE

A WIDE WOOD PANEL IS ALMOST IMPOSSIBLE TO KEEP FROM CUPPING ON THE SAP SIDE.

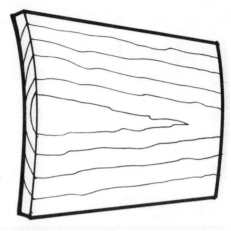

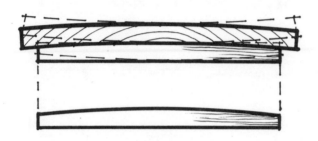

MY DAD OVERCAME THIS BY SCREWING CURVED CLEATS TO THE BACKSIDE OF THE PANEL. THEY PULL THE PANEL TO A CURVE OPPOSITE THE ORIGINAL CUP.

THE BOARD WILL TRY TO CUP AGAIN BUT WILL GO JUST SO FAR. HOW MUCH TO CURVE THE CLEATS IS PRETTY MUCH GUESS WORK. A PIECE OF ANGLE IRON SCREWED TO THE BACK WLLL ALSO WORK.

NOTHING DISTURBS A ROOM MORE THAN A SAGGING BOOK SHELF. 3/4" STOCK SHOULD BE NO LONGER THAN 30".

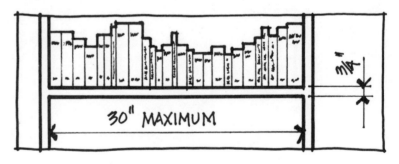

30" MAXIMUM

DRAFTING:
TIPS AND TRICKS ON DRAWING AND DESIGNING HOUSE PLANS

BOB SYVANEN

OVER 250 ILLUSTRATIONS

THE LONGER THE SHELF, THE THICKER IT SHOULD BE UNLESS IT HAS SUPPORT FRONT AND BACK.

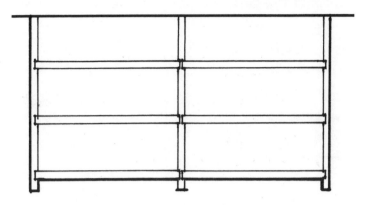

STRONG, NEAT SHELVES CAN BE MADE WITH GROOVED SIDE PANELS.

LINING UP THE ADJACENT GROOVE IS ALWAYS A TOUGH JOB.

THIS JIG MADE OF 3/4" PLYWOOD MAKES EASY WORK OF IT. MAKE THE SPACERS THINNER THAN THE BOARD BEING GROOVED OR SHIM THE BOARD WITH A PIECE OF CARDBOARD TO INSURE A GOOD GRIP BY THE JIG.

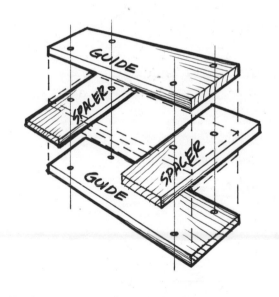

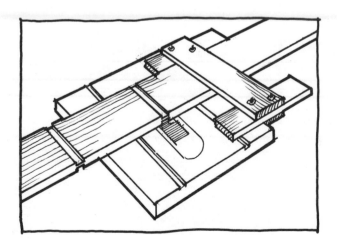

USE THE EDGE OF THE TABLE TO RUN THE GUIDE AGAINST. FLIP THE WHOLE BUSINESS OVER AND CUT THE MATCHING GROOVE. MAKE SURE THE JIG IS CUTTING SQUARE.

MISCELLANEOUS

I LIKE TO SET UP A WORK TABLE AS SOON AS POSSIBLE. FIND A LONG FREE SPACE WHERE A 2×10 × 14'-0" PLANK CAN BE SET UP. 37" HIGH SUITS ME.

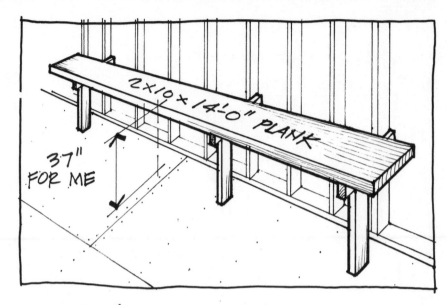

37" FOR ME

2×10 × 14'-0" PLANK

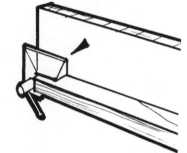

A CARPENTER'S VISE AT ONE END IS A GREAT HELP.

WHEN EASING CORNERS WITH A BLOCK PLANE, RUN THE PLANE, IN ONE STROKE, THE LENGTH OF THE BOARD. THIS WILL GIVE A NICE EVEN CHAMFER.

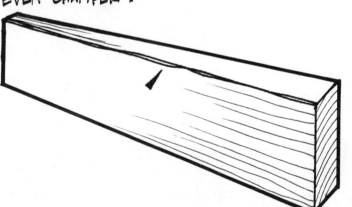

SHORT CHOPPY STROKES WILL LEAVE AN IRREGULAR CHAMFER.

THE SAME APPLIES FOR ANY PLANING: LONG SMOOTH STROKES MAKE FOR A SMOOTH JOB. KEEP THE AREA ALONG THE WORK TABLE CLEAR SO THAT THERE IS NO STOPPING AS YOU WALK WITH THE PLANE FROM END TO END.

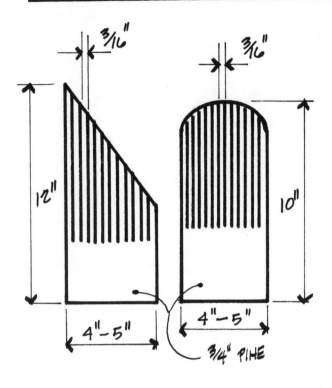

3/16"

3/16"

12"

10"

4" – 5"

4" – 5"

3/4" PINE

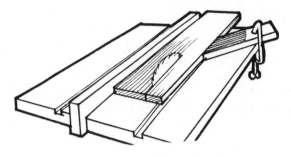

SPRING BLOCKS CLAMPED TO THE TABLE SAW TOP KEEP STOCK BEING RIPPED AGAINST THE RIP FENCE WHILE YOU CONCENTRATE ON PUSHING THE PIECE THROUGH.

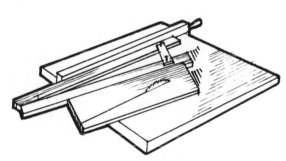

A QUICKIE TAPER JIG CAN BE MADE WITH TWO 1 X 3 X 20" OR 24" PIECES OF STRAIGHT STOCK, A HINGE OR 1 X 2 BLOCK AT ONE END, AND A 1 X 2 TACKED AT THE OTHER TO HOLD THE ANGLE. ADD A STOP AT THE BOTTOM OF ONE LEG AND IT'S READY TO GO.

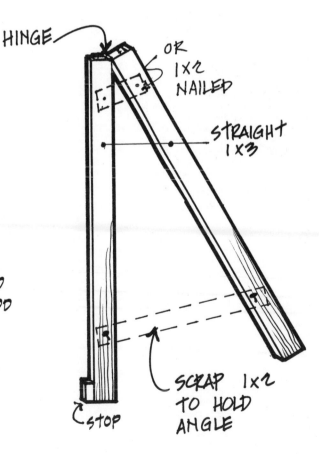

HINGE

OR 1 X 2 NAILED

STRAIGHT 1 X 3

SCRAP 1 X 2 TO HOLD ANGLE

STOP

MISCELLANEOUS

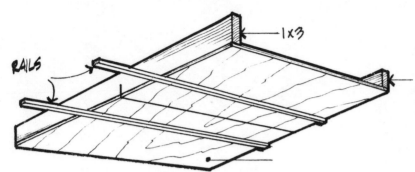

RAILS

1x3

THIS IS A HANDY JIG FOR CUTTING WIDE BOARDS, NICE FOR CABINET WORK. SPRAY THE RAILS WITH SILICONE FOR EASY SLIDING.

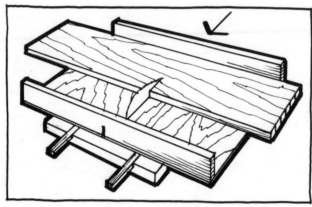

THIS IS A NICE JIG FOR MAKING RAISED PANELS ON THE TABLE SAW. SPRAY THE RIP FENCE FOR EASY SLIDING.

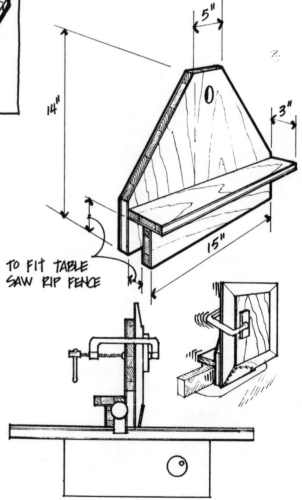

5"

3"

14"

15"

TO FIT TABLE SAW RIP FENCE

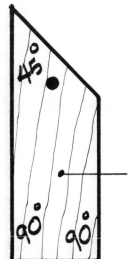

3/4" x 8" x 16" PLYWOOD

THE MITER GAUGE ON A TABLE SAW CAN BE QUICKLY CHECKED IF A PATTERN IS KEPT HANDY.

SHARPENING STONES SHOULD BE KEPT IN A BOX AND WIPED CLEAN AFTER EVERY SHARPENING. A QUICK WAY TO MAKE A BOX IS TO PUT THE STONE ON A PIECE OF 3/4" PINE (LARGER THAN THE STONE) AND NAIL SOME STRIPS, ONE HALF THE THICKNESS (PLUS A TAD) OF THE STONE ON THE BOARD ALL AROUND THE STONE. MAKE TWO LIKE THIS AND TRIM TO SIZE ON THE TABLE SAW. KEEP A LITTLE CLEARANCE BETWEEN THE STRIPS AND THE STONE.

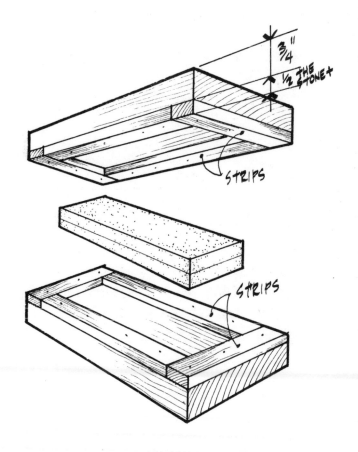

3/4"

1/2 THE STONE +

STRIPS

STRIPS

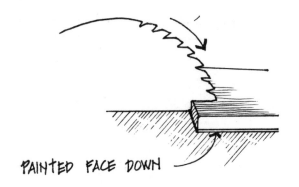

PAINTED FACE DOWN

WHEN CUTTING WOOD PAINTED ON ONE SIDE, PUT THE PAINTED SIDE DOWN SO THE BLADE WONT PULL AS QUICKLY.

MISCELLANEOUS

GLUEING PIECES OF WOOD TOGETHER SOUNDS SIMPLE, BUT THERE ARE SOME BASIC TECHNIQUES THAT CAN HELP.

WHEN GLUEING BOARDS EDGE TO EDGE FOR A BENCH OR TABLE TOP, THE SAP AND HEART FACES SHOULD ALTERNATE TO MAINTAIN A FLAT SURFACE.

HEART   SAP   HEART

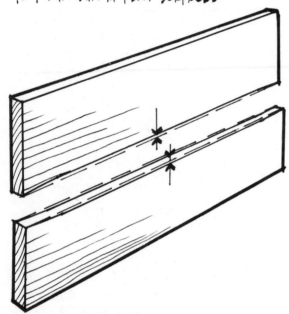

THEY SHOULD BE PLANED, SLIGHTLY, AT THE CENTER, TAPERING TO NO PLANING AT THE ENDS. THE EDGE OF THE BOARDS WILL SHRINK MORE THAN THE MIDDLE SO EVERYTHING WILL EQUALIZE. IF THIS PLANING ISN'T DONE, THE ENDS WILL SHRINK AND SPLIT. OBSERVE HOW BOARDS SPLIT AT THE ENDS.

TWO BOARDS GLUED FACE TO FACE SHOULD HAVE THE SAP FACES TOGETHER AND THE GRAIN RUNNING IN THE SAME DIRECTION, NOT CROSS GRAINED LIKE PLYWOOD.

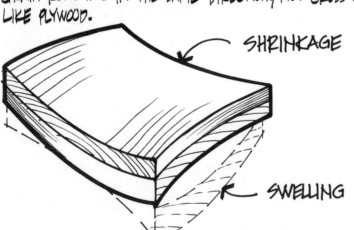

SHRINKAGE

SWELLING

IF TWO BOARDS ARE CROSS GRAINED, THE CHANCES FOR WARPING IS GREAT. THREE BOARDS CAN BE CROSS GRAINED.

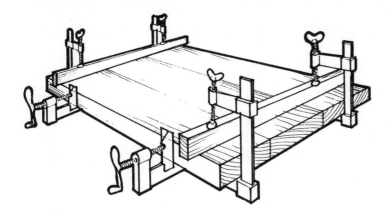

THIS IS THE BEST WAY TO CLAMP EDGE-GLUED BOARDS.

IF BAR CLAMPS ARE TOO SHORT THEY CAN BE USED IN COMBINATION.

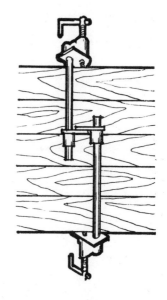

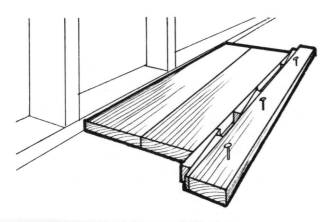

HERE IS A QUICKIE WAY TO EDGE-GLUE BOARDS. SOME WEIGHTS ON THE TOP WILL KEEP THE BOARDS FROM BUCKLING.

MITERS ARE TOUGH TO CLAMP, BUT TWO BLOCKS TACKED ON THE OUTSIDE WILL MAKE CLAMPING POSSIBLE.

IF THERE IS WATER AVAILABLE, THIS IS A GOOD WAY TO GET SOME PRESSURE FOR FACE GLUEING.

MISCELLANEOUS

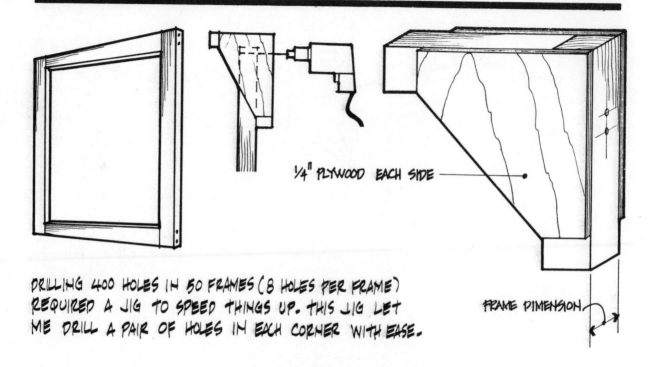

¼" PLYWOOD EACH SIDE

FRAME DIMENSION

DRILLING 400 HOLES IN 50 FRAMES (8 HOLES PER FRAME)
REQUIRED A JIG TO SPEED THINGS UP. THIS JIG LET
ME DRILL A PAIR OF HOLES IN EACH CORNER WITH EASE.

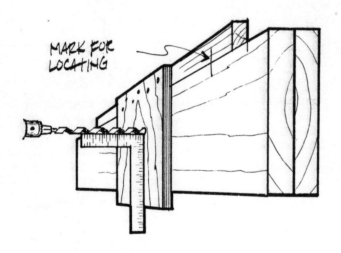

MARK FOR
LOCATING

TWENTY TWO BEAMS TWENTY INCHES
THICK HAD TO BE DRILLED FROM EACH
SIDE. AND THE HOLES HAD TO MEET.
I MADE A JIG TO LINE UP THE
DRILLING AND IT WORKED GREAT.

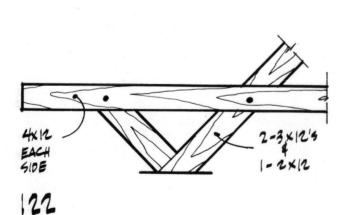

4X12
EACH
SIDE

2-3X12'S
&
1-2X12

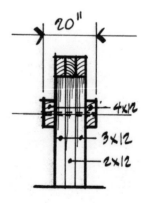

20"

4X12

3X12

2X12

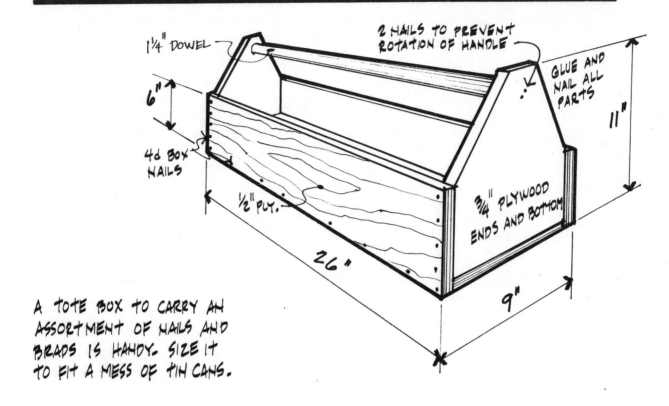

1¼" DOWEL

2 NAILS TO PREVENT ROTATION OF HANDLE

GLUE AND NAIL ALL PARTS

6"

11"

4d BOX NAILS

½" PLY.

¾" PLYWOOD ENDS AND BOTTOM

26"

9"

A TOTE BOX TO CARRY AN ASSORTMENT OF NAILS AND BRADS IS HANDY. SIZE IT TO FIT A MESS OF TIN CANS.

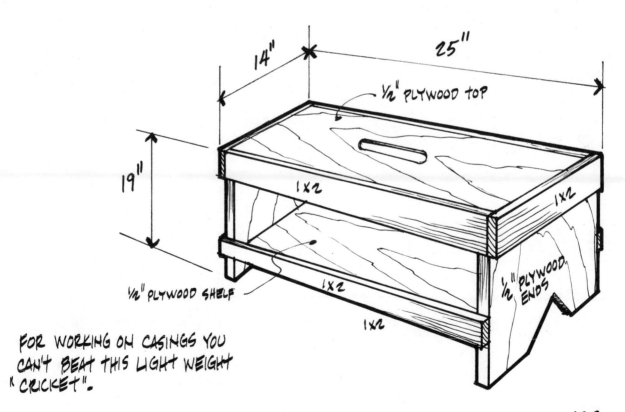

14"

25"

½" PLYWOOD TOP

19"

1X2

1X2

½" PLYWOOD SHELF

1X2

½" PLYWOOD ENDS

1X2

FOR WORKING ON CASINGS YOU CAN'T BEAT THIS LIGHT WEIGHT "CRICKET".

MISCELLANEOUS

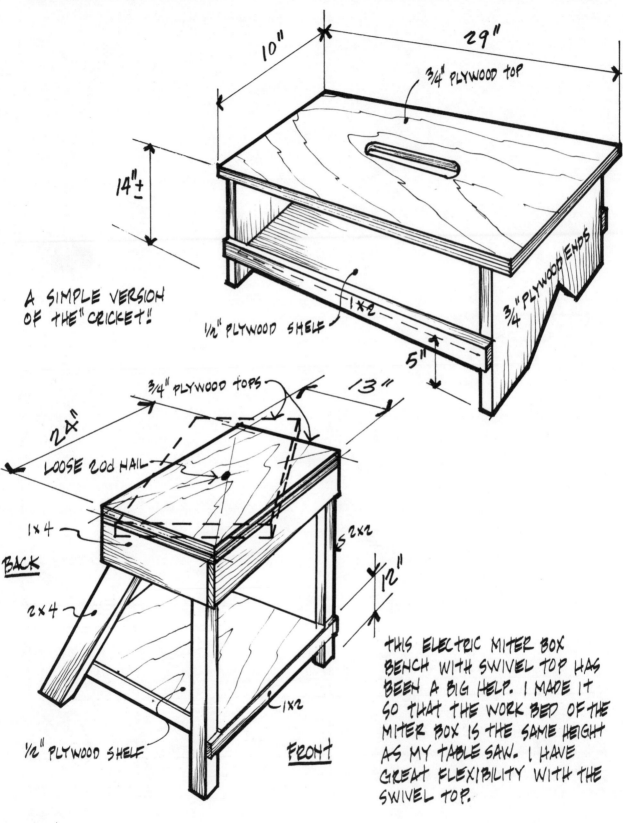

10"

29"

3/4" PLYWOOD TOP

14"±

A SIMPLE VERSION OF THE "CRICKET!"

1/2" PLYWOOD SHELF

1×2

3/4" PLYWOOD ENDS

5"

3/4" PLYWOOD TOPS

13"

24"

LOOSE 20d NAIL

1×4

2×2

BACK

2×4

12"

1/2" PLYWOOD SHELF

1×2

FRONT

THIS ELECTRIC MITER BOX BENCH WITH SWIVEL TOP HAS BEEN A BIG HELP. I MADE IT SO THAT THE WORK BED OF THE MITER BOX IS THE SAME HEIGHT AS MY TABLE SAW. I HAVE GREAT FLEXIBILITY WITH THE SWIVEL TOP.

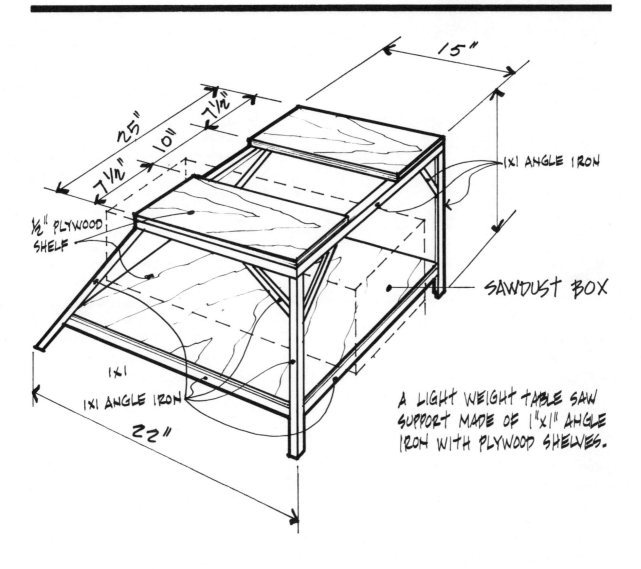

15"

25"

7½"

10"

7½"

1X1 ANGLE IRON

½" PLYWOOD SHELF

SAWDUST BOX

1X1

1X1 ANGLE IRON

22"

A LIGHT WEIGHT TABLE SAW SUPPORT MADE OF 1"X1" ANGLE IRON WITH PLYWOOD SHELVES.

A SAWDUST BOX UNDER THE TABLE SAW KEEPS THE AREA CLEAN BUT WHEN THE ARBOR NUT FALLS INTO THE SAWDUST, IT IS ALMOST IMPOSSIBLE TO FIND. KEEP A MAGNET HANDY AND FISHING OUT METAL IN THE SAWDUST BOX IS A SNAP.

SAW DUST

MISCELLANEOUS

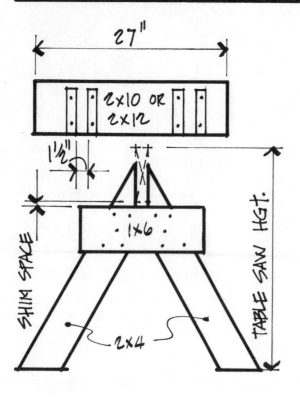

27"

2×10 OR
2×12

1½"

SHIM SPACE

·1×6·

2×4

TABLE SAW HGT.

THIS IS A QUICKIE WORK SUPPORT
FOR THE TABLE SAW. IT COMES
APART, IS EASILY STORED AND
ADJUSTS FOR HEIGHT.

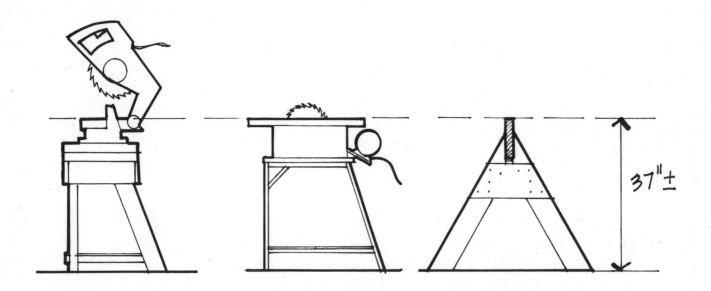

37"±

I MADE THE HEIGHT OF ALL MY EQUIPMENT SO THAT I COULD USE THE PIECES WITH
EACH OTHER.

CARPENTRY:

SOME TRICKS OF THE TRADE .....FROM AN OLD-STYLE CARPENTER

BOB SYVANEN

OVER 400 ILLUSTRATIONS
ILLUSTRATED BY MALCOLM WELLS

DRAFTING:

TIPS AND TRICKS ON DRAWING AND DESIGNING HOUSE PLANS

BOB SYVANEN

OVER 250 ILLUSTRATIONS

INTERIOR FINISH:

MORE TRICKS OF THE TRADE .....FROM AN OLD-STYLE CARPENTER

BOB SYVANEN

OVER 500 ILLUSTRATIONS

With more than 30 years of experience as a carpenter and architect, Bob Syvanen really knows the tricks of the trade. Each of his three books will solve common problems that crop up unexpectedly and will be an invaluable tool for both the beginner and the expert!

"These are good little books with the sort of practical information most similar volumes leave out."

—The Charlotte Observer

"...valuable supplements to basic building guides."

—Publishers Weekly

"These no-frills paperbacks...should be a must for any beginner and probably can teach even experienced builders a thing or two."

—The Sacramento Union

Order Today!

- -

The East Woods Press, 429 East Blvd., Charlotte, N.C. 28203 (704) 334-0897

Please send:

_____ **Carpentry $7.95**

_____ **Drafting $7.95**

_____ **Interior Finish $7.95**

Please add $1.30 for postage for the first book plus $.50 for each additional book.

Send to:

_____

_____

_____

_____

My Check for _____ is enclosed.

Charge my Visa _____ Exp. date _____

Mastercard _____ Exp. date _____

Signature _____

East Woods Press Books

Backcountry Cooking
Berkshire Trails for Walking & Ski Touring
Campfire Chillers
Canoeing the Jersey Pine Barrens
Carolina Seashells
Carpentry: Some Tricks of the Trade from an Old-Style Carpenter
Catfish Cookbook, The
Complete Guide to Backpacking in Canada
Drafting: Tips and Tricks on Drawing and Designing House Plans
Exploring Nova Scotia
Free Attractions, USA
Free Campgrounds, USA
Fructose Cookbook, The
Grand Strand: An Uncommon Guide to Myrtle Beach, The
Healthy Trail Food Book, The
Hiking from Inn to Inn
Hiking Virginia's National Forests
Honky Tonkin': Travel Guide to American Music
Hosteling USA, Revised Edition
Inside Outward Bound
Interior Finish: More Tricks of the Trade
   from an Old-Style Carpenter
Just Folks: Visitin' with Carolina People
Kays Gary, Columnist
Living Land: An Outdoor Guide to North Carolina, The
Making Food Beautiful
Maine Coast: A Nature Lover's Guide, The
New England Guest House Book, The
New England: Off the Beaten Path
Parent Power!
   A Common-Sense Approach to Raising Your Children In The Eighties
Race, Rock and Religion
Rocky Mountain National Park Hiking Trails
Sea Islands of the South
Southern Guest House Book, The
Southern Rock: A Climber's Guide to the South
Steppin' Out: A Guide to Live Music in Manhattan
Sweets Without Guilt
Tennessee Trails
Train Trips: Exploring America by Rail
Trout Fishing the Southern Appalachians
Vacationer's Guide to Orlando and Central Florida, A
Walks in the Catskills
Walks in the Great Smokies
Walks with Nature in Rocky Mountain National Park
Whitewater Rafting in Eastern America
Wild Places of the South
Woman's Journey, A
You Can't Live on Radishes

Order from:
The East Woods Press
429 East Blvd.
Charlotte, NC 28203